T0262592

The Tangle of Science

The Tangle of Science

Reliability Beyond Method,
Rigour, and Objectivity

Nancy Cartwright, Jeremy Hardie,
Eleonora Montuschi, Matthew Soleiman,
and Ann C. Thresher

OXFORD
UNIVERSITY PRESS

OXFORD
UNIVERSITY PRESS

Great Clarendon Street, Oxford, OX2 6DP,
United Kingdom

Oxford University Press is a department of the University of Oxford.
It furthers the University's objective of excellence in research, scholarship,
and education by publishing worldwide. Oxford is a registered trade mark of
Oxford University Press in the UK and in certain other countries

Published in the United States of America by Oxford University Press
198 Madison Avenue, New York, NY 10016, United States of America

British Library Cataloguing in Publication Data
Data available

Library of Congress Control Number: 2022939675

ISBN 978-0-19-886634-3

DOI: 10.1093/oso/9780198866343.001.0001

Printed and bound in the UK by
Clays Ltd, Elcograf S.p.A.

Links to third party websites are provided by Oxford in good faith and
for information only. Oxford disclaims any responsibility for the materials
contained in any third party website referenced in this work.

An African Jacana Bird building its floating nest.
Source: photogallet/DepositPhotos.

Contents

Acknowledgements

This book is the result of a research programme that the authors have developed together and in dialogue with a number of colleagues who have greatly contributed to the advancement of the contemporary field of philosophy of science.

A huge debt goes to Hasok Chang. His work on temperature is one of two key sources we draw on for our illustration of the tangle in Chapter 5. More importantly, *The Tangle of Science* is a study in tune with the movement for the philosophy of science-in-practice. Chang has led the movement for this kind of philosophy. You will also see echoes of his monumental work on pragmatic theory of truth.

Many ideas for this book have evolved in conversation with Sharon Crasnow and with Alison Wylie and in parallel with their own work. What we offer here is a general framework that we believe dovetails with their specific proposals. We are very grateful for all their help and for their contributions.

We also owe a big debt to Peter Vickers who read the whole manuscript and offered much helpful advice (some of which we took!) and to Nicola Craigs and Giulia Gandolfi who provided invaluable help with the production of the manuscript. We are also grateful to Adrian Williams who advised on Meccano, Sindhuja Bhakthavatsalam who taught Nancy Cartwright how important it is to relate endeavours to purposes (see, for instance, Bhakthavatsalam 2019), Cathy Gere for her comments on parts of the book, Craig Callender for his advice on gravitational waves, and Anna Alexandrova for inspiration. Lucy Charlton provided many of the illustrations.

Section 1.5 draws heavily on Nancy Cartwright's (2020) *Aristotelian Society* paper, 'Why trust science? Reliability, particularity and the tangle of science' and the discussion of Edna-Ullman Margalit's *Out of the Cave* in Section 4.5, from Cartwright (2014).

A version of Sections 3.2.2, 3.2.3, and 3.4.1 can be found in Eleonora Montuschi's 'Finding a context for objectivity', *Synthese* 199 (2021).

Much of Cartwright's and Montuschi's research for this book was supported by Knowledge for Use (K4U). The K4U project received funding from the European Research Council (ERC) under the European Union's Horizon 2020 research and innovation programme (grant agreement No 667526 K4U).

The content reflects only the authors' view. The ERC is not responsible for any use that may be made of the information it contains. Cartwright, Soleiman, and Thresher all received support from the summer research fund at the University of California at San Diego. Soleiman also received support from UC San Diego's Science Studies Program.

List of Figures

Introduction

In Brief. Science creates multitudes of products that do what you expect of them: theories, models, concepts, experiments, measurement procedures, technological devices, social policy evaluations, social and archaeological narratives, approximation techniques and many, many more. What underwrites their reliability? We group the standard answers under the headings 'scientific method', 'rigour', and 'objectivity'. Though they have a role to play, these 'usual suspects', we argue, are not up to the job of supporting the effectiveness of these products, whether individually or in combination. What, then, can provide such support? It is not just these usual suspects or anything like them. Rather, the reliability of any one product in science generally rests on a vast, amorphous network of other heterogeneous scientific products, usually unnoted and undersung, that woven together, as in a Jacana bird's nest, help provide the secure support it needs. This is the *tangle of science* referred to in our title.

In more detail. We put people on the moon and robots on Mars, we teach our children that the earth is not flat but roughly spherical and that water conducts electricity, we happily submit to laser cataract surgery, and we invest heavily in studying the genetic structure of viruses. Rightly so—you can fly round the earth to see its shape for yourself, people get killed by using hairdryers in the bath, laser cataract surgery goes wrong only once in around 5,000 times, and gene-based research has helped to develop vaccines that prevent deadly illnesses. Clearly much of what science offers can do what you expect of it. But what is it about science that makes its products reliable?

Bodies like the US National Academy of Sciences, the UK's Science Council, and the American Physical Society tell you that the good thing about science is that it is 'systematic', it is 'testable', and it uses 'empirical evidence'. This puts a lot of pressure on these words—just how are they to be understood?

How systematic is systematic, and systematic in what sense? Is physics more scientific than anatomy or geology because it can be systematised via sets of equations? And what, after all, should count as *evidence* for or against

The Tangle of Science: Reliability Beyond Method, Rigour, and Objectivity. Nancy Cartwright, Jeremy Hardie, Eleonora Montuschi, Matthew Soleiman, and Ann C. Thresher, Oxford University Press. © Nancy Cartwright, Jeremy Hardie, Eleonora Montuschi, Matthew Soleiman, and Ann C. Thresher 2022. DOI: 10.1093/oso/9780198866343.003.0001

something, and is evidence enough or must it be of the right kind and variety? Do we need science to make *novel* predictions—to predict facts we didn't already know—or is it ok if it just accommodates some of what is well-known? And does it matter how precise the predictions are?

There is a vast amount of both philosophical and scientific work on these questions. The traditionally cited reasons for the reliability of science are the meat of Part I of this book. We organise these under three headings: science uses *the scientific method*, it is *rigorous*, and it is *objective*. Part I rehearses the strengths and the limitations of each, arguing that these three pillars cannot generally deliver the degree of reliability that we all depend on in using the products of science, nor can they secure scientific knowledge. We argue:

- *There is no scientific method.* Anything general enough to cover the vast array of what is normally characterised as science is too vague to do serious work, as has been often noted. Nor should there be. Demarcating what is and is not scientific method or marking out sets of peculiarly *scientific* methods is a mistake, contrary to the openness of inquiry that makes for credibility and for scientific advance. Worse, the hunt for scientific method is standardly tied to the task of *theory confirmation*, ignoring how the reliability of all the other products essential to science is to be secured.
- *Rigour is altogether the wrong notion.* It is a virtue but it cannot deliver much. What can be established rigorously is narrow in scope. Nor do heaps of rigorous results add up to solid support, as many hope.
- *The usual notion of objectivity—the correct application of pre-agreed procedures for pre-agreed ends—is not good enough for science.* The kind of objectivity that is needed requires that both the right procedures and the right ends be found—in tandem, case-by-case.

Beyond that, there are the usual, more general sources of worry about how much stable reliability we can expect from science: underdetermination, pessimistic meta-induction, new and old problems of induction, and well-known problems with confirmation theories. We describe these briefly in Section 1.2. Despite all these limitations, we agree that the traditional reasons do speak in favour of reliability. But they are not enough to make it firm.

There is no need for despair though. There is far more to the inner workings of science that can yield reliability than what these traditional pillars point to. But to get it, the enterprise needs to be re-conceived. The standard approach to the question 'What makes science so trustworthy?', both in philosophy and in the detailed methodologies of the various sciences, is to ask 'What constitutes *confirmation* for scientific *claims*?'. We urge a dramatic broadening of focus: from concentrating on claims and theories to focusing on *all* the products of science and not on their *truth* but on their *reliability* and on the *purposes* they can reliably serve. Most of these products (e.g. methods, measures, experiments, classification schemes, approximation techniques, and technological devices) are not candidates for truth in the first place. Yet it is vitally important that they be able to do what is expected of them.

So this book does not concentrate just on claims but rather on the vast array of science elsewhere, on all the products that science creates: in models; measurement definitions, procedures, and instruments; concept development and validation; data collection, analysis, and curation; experimental and non-experimental studies; statistical techniques; methods of approximation; case studies; narratives; etc.; etc.; etc. And especially on how each of these must itself be reliable if the reliability of any one of them is to be vouchsafed— where this includes the highly prized claims of science, its theories and its laws.

When we urge a focus on all the products of science, we use the word 'product' purposely. We want you to think of products, the kinds of things one finds on shop shelves. These have been envisioned, developed, created, assembled, and tested by one conglomerate of actors and put on the shelf to be taken down for use by different actors with different ideas and practices for employing them in their own endeavours.

Many products on shop shelves come with instructions for use: 'You can bake good cakes with this soft flour but not good pasta'; 'Take 2 tablets twice a day, best with food'; 'Store in a cool dry place'; or 'Use [Genie Crafts Cloth Plaster Wrap Rolls] to make keepsakes, pregnancy belly casting, masks, science projects, 3D sculptures, science projects, etc.'[1] But new users find new uses: 'We've used this [Swiss Army knife]...to prepare tinder, fashion snare triggers, tighten screws, strip wires, file corroded wires, open cans, repair toys, make an alcohol stove from a tuna can, and clear shooting lanes' (Barlow 2017). Users also share information beyond that in the official guides about what you can and cannot do with the product: 'Do not try to use this product [Genie Crafts Cloth Plaster Wrap Rolls] for machete [*sic*] as it heats up while it dries and so if you use a balloon as the base it will pop!'[2]

---

[1] See https://www.amazon.co.uk/4-Pack-Plaster-Crafts-Projects-Inches/dp/B07PM3XL4C (accessed 13 October 2020).

[2] Elwin in customer product reviews: see https://www.amazon.co.uk/dp/B07PM3XL4C#customerReviews (accessed 13 October 2020).

And there are general shared understandings: you don't expect your mobile phone to work in the middle of a cave or if you drop it in the bath.

The same is true for the products of science. There are only a few explicit users' manuals but a lot of implicit and shared understanding—as with Andrea Woody's (2015) 'communal norms' that we introduce in Section 4.5. Scientists, and ordinary users as well, learn what their products can and cannot be used for in combination with others. Just as soft flour can be used with different combinations of ingredients to make a Victoria sponge or play dough or to thicken Bechamel sauce but there doesn't seem to be any combination that turns it into good pasta, so too with science, where learning what can be used with what for what ends is standard procedure.

For just one among millions of examples, consider the use of continuum versus particulate models for flow. Here is one discussion among the myriad you can find on this topic, which we quote at length to underline the detail of the lessons that are recorded.

> Some researchers have used a continuum approximation to obtain a version of hydrodynamics for granular media. The strategy is sometimes useful—for example, in treating energy transport in highly excited granular matter. However, the continuum approach can be problematic. When the degree of excitation is low, gravity often produces persistent contacts between particles, so their interactions are not limited to the isolated collisions that are assumed in justifying a continuum approximation. Also, if you shake a granular medium to excite it, the energy subsequently decays rapidly but unevenly due to huge numbers of inelastic collisions. This collision process leads to inhomogeneities in the local kinetic energy on scales only a few times larger than the particle size. The distribution of internal stresses in both the static and dynamic states is highly nonuniform, and the stresses are transmitted along linear chains of particles, in contrast to the situation in ordinary solids. Researchers have devoted much attention to understanding such effects. Because the inhomogeneities are so prominent, it is difficult to treat the network of forces between interacting particles accurately using continuum methods, although one promising research effort is to determine whether there is some length scale above which spatial averaging makes sense. (Gollub 2003, 10)

The standard answers to the more usual question, 'What constitutes confirmation for scientific claims?', are formal theories of confirmation, detailed sets of procedures for conducting research that one learns in becoming a practising scientist, or informal criteria like unifying power, prediction of novel results, and simplicity. We maintain that the emphasis should be elsewhere. What goes much further to securing the reliability of any product of science is the great *tangle* of other scientific products that back it up. By *tangle* we mean the rich interwoven net of scientific creations that constrain and support each other—the concurrent, mutually feeding back and forth, developing network of ideas, concepts, theories, experiments, measures, bridge principles, models, methods of inference, research traditions, data and narratives, etc. etc. that make up a scientific endeavour, with its long tentacles out into other similarly rich tangles that it rests on and that can in turn rest on ingredients from it. The focus should not be on how to warrant the truth of scientific claims but rather on how to *accredit* this *tangle of work*—on how to gain assurance that the body of work that supports the reliability of a scientific product is up to the job.

Claims, especially law claims and theories, function in many ways. They may, for instance, be a help at unification or perhaps play some special role in explanation. But our concern is reliability. From that point of view claims are in the same boat as all the other products of science. When it comes to vouchsafing the reliability of any of these products, law claims and theories usually play a part, but just a part, along with a great many other scientific products. And they need a tangle of other products behind them to vouch-safe their own reliability, whatever the job in view is, including telling the truth.

Even when you do wish to focus narrowly on claims and their truth, what makes it likely that a claim can do what you want it to—in this case, tell the truth about the world—is no different from what makes for reliability of any other product of science. The reliability of the claim is derivative from features of the complex tangle that supports it. No claim is an island: the meaning of any concept, equation, or principle depends on the vast network of other scientific work in which it is embedded. And no claim stands or falls by just some fixed body of evidence. Facts do not come with a label: 'I am evidence for this claim'. Facts become evidence for this or that hypothesis only relative to a vast panoply of other claims and practices, models, theories, data, and concepts.

Consider Maria Elena Di Bucchianico's (2009) story of two warring camps in the field of high-temperature superconductivity. One took the explanatory mechanism to be phonons; the other, magnetic modes. In 2002, new methods (ARPES—Angle-resolved photoemission spectroscopy) showed a 'kink' in the dispersion curve of reflected photoelectrons. Both camps agreed that the data were correct. But they had wildly different interpretations of it due to the great number of differing assumptions and practices they were also committed to. In fact, both claimed the result to be in their favour and against the other. As Di Bucchianico reports, '[E]ach of [the] two warring camps claim to account for the same evidence, meanwhile claiming that this evidence is incompatible with the opposing theory' (12).

So, you may be impressed by a new result—say a wonderfully precise novel prediction born out in a carefully conducted experiment—and take it to confirm your favoured theory. But whether it does what you think it does depends on a host of other assumptions being true, other experiments having been well conducted, other concepts being true to the world, and so forth. Of course, not every assumption has to be right, not every experiment in the tangle has to have been well conducted: a good thing about good science is that methods, assumptions, concepts, and so on are all multiply supported in a myriad of different ways so that one line of defence may crumble, or even many perhaps, without the whole edifice falling. But enough must be right, and in some right mix.

Even with endless time to draw a list, though, you cannot say what your hypothesis really rests on. When you say that this exciting new piece of evidence confirms it, you count on enough of these tangles upon tangles being right enough in a good enough mix to ensure that the evidence plays the role you suppose, even though you cannot delineate just which mix of which strands from which tangles. You rely on some not very clearly specified body of scientific work being credible enough to do the job. That is why we say that warranting that a scientific claim can be relied on to tell the truth is just like warranting that the laser will emit a precise wavelength of cool ultraviolet light that will not burn your eye: it involves *accrediting a vast tangle of scientific work that backs it up.*

We liken the vast, dense tangle of scientific work that supports a scientific principle or product to the nest that a Jacana bird builds to support its precious eggs floating on water in a shallow lake in Africa. Twigs, stems, and leaves are carefully interwoven, each supporting the others to create a secure nest. No single piece of the nest is the foundation for the rest, nor would the whole thing fail should one or two twigs be pulled away. Remove enough, however, or build it badly, and the entire structure may collapse, or you may end up with gaps that the eggs can fall through.

Not all bodies of background scientific work are good enough to warrant the expectation that a product can be relied on to serve the purposes you want it to. Following a suggestion of Alison Wylie, we call those that are up to this job '*virtuous* tangles', and we offer three general features that together, we claim, are crucial for a body of work to be virtuous: the body of work should be *rich* (there are lots of pieces and of many different kinds), the pieces are *entangled* (the pieces relate to each other and to the product in a variety of different ways, and these are the 'right' ways for the job), and *long-tailed* (the pieces figure in support tangles for other scientific products in other domains that are succeeding at other difficult jobs, including successful interactions with the empirical world).

Our claim is both descriptive and normative—this is what generally is there when a scientific product works reliably, and it is what you should demand if you are to bet on it. What we offer is not like confirmation theory, which purports to tell you how likely it is that your conclusion is correct. Nor can you use our three criteria as a checklist to tick off in order to decide how good a bet the product is. Science is not like that, we claim, despite the love of 'p values' and confidence intervals in the social and medical sciences, the trust-generating rigour of mathematical approaches in both the natural and social sciences, and the precision of experiments from virology to gravity wave detection. As we argue in Chapter 2, what can be achieved rigorously is of extremely limited use, and even what counts as evidence for what depends on whole bodies of often hidden work in the background. We close with a cautionary lesson: no matter how good a tangle looks, it may well let you down. We aim to show what makes science more reliable but not how to be certain about it—that's not humanly possible.

The three criteria we offer are general, abstract. We explain what we mean by them and give illustrations, both brief illustrations in Chapter 4 and more elaborate ones in Chapters 5 and 6. We can say a lot about them, especially once we get down to brass tacks, case by case. But we cannot define them beyond what we explain in Chapter 4, nor say anything operationalisable. That, we claim, is characteristic of science itself. There are pockets of great rigour, which are an important part of the enterprise. But accrediting the reliability of scientific products cannot be done 'by the book'. This does not mean that there is not a lot to be said, case by case. But what is to be said will have to come from deep within the details of the science itself and the context of use. Nor does it mean that there are no rules of thumb. To the contrary. If you are depending on a generic—like 'People respond to incentives' or 'Limestone wears its flaws on its face'—you had better consider the cost of mistakenly assuming the generic holds in the case at hand versus the cost of local investigation to see what might get in its way. Or, if you are relying on measurements made by real people, you should insist on a study of inter-rater consistency.

So, we tell you that for reliability you should expect a 'virtuous' tangle: the 'right' kind and variety of ingredients entangled in the 'right' ways and connected with the 'right' bodies of work elsewhere. But we do not offer a prescription on what counts as 'right'. Nor, we think, can anyone else, beyond vague generalities and product-, purpose-, and context-relative rules of thumb. We offer an account of what science does, beyond exhibiting virtues like objectivity, rigour, and adherence to special scientific methods, that makes its products so reliable—it produces virtuous tangles. But we cannot provide you with a recipe for telling if a tangle is virtuous. That is a case-by-case matter, settled by case-by-case specifics.[3]

Note that the idea of the tangle cuts two ways.

On the one hand, a virtuous tangle can increase reliability, especially when the job you want the scientific product to do is narrow and precisely characterised. We explain how this can happen in Part II where we say more about the tangle and provide a variety of illustrations of how it works in the natural and social sciences.

On the other hand, the tangle threatens truth. In the last few decades, philosophers of science have been obsessed by a general reason supplied by the history of science not to suppose that scientific knowledge is secure: 'the pessimistic meta-induction'. There has been a string of theories that were thought well-confirmed in the past but that are no longer taken to be true. Why suppose any better fate for those just now when we live?

We agree that there is a good reason to expect the overthrow of 'well-established' scientific principles. But for a different reason than the pessimistic meta-induction or the problems of induction, underdetermination, and the like. Rather, it is because these are seldom as well-established as it might have seemed. You can often, post hoc, find a number of crucial gaps in past support networks that never were filled satisfactorily—the tangle wasn't good enough to do the job well. But beyond specific flaws, there is a more general reason: the tangle that supports any of our scientific principles, no matter how dense and large, is after all like the tangle of twigs, grass, stems, and leaves that supports the African Jacana bird's eggs. It ultimately floats on water, as Otto Neurath ([1921] 1973) taught in his famous metaphor that we are like sailors rebuilding our ships at sea, never able to put in to dry dock to build from firm foundations.

Science's estimates of how secure a claim is are made using science's views at the time about what else is known, what methods are effective, what concepts are valid, what narratives and models are plausible, what data is correctly categorised, and the like. If you really want to estimate how

---

[3] There is a spate of work recently concerned with how to tell when scientific claims are likely to be trustworthy and when not (e.g. Conway and Orekes 2011; Vickers 2022). For instance, has the defence of the claim been peer reviewed, does it report a result favourable to a commercial enterprise that paid for the study, are the authors actually expert in the details it takes to defend the claim, does it have the backing of the overwhelming majority of the relevant scientific community, has it stood the test of time? Some of these suggestions may well double as symptoms of reliability. This seems plausible and is well worth investigating but it is not an issue we explore here.

well-established a claim is, as opposed to 'how well-established for the nonce' or 'how well-established relative to this body of knowledge and practice', then the image of the floating nest should remind you that these estimates all need to be deflated.

What holds for truth holds equally for reliability. It too depends on tangles upon tangles, none ever rooted firmly. The products of science, even when well-warranted by the best science of the time, will never be totally reliable. But you can try to identify and encourage the features that make it more so. That is what we set out to do. We aim to identify and elucidate features commonly found in science as it is practised that are conducive to the production of reliable outputs.

Situating Our Work

Our focus on the products of science and on what makes them reliable complements accounts from those such as historian Naomi Oreskes (2019), sociologist Karin Knorr Cetina (2009), and philosophers like Philip Kitcher (2003) and Helen Longino (1990, 2002) on the social structures and institutions that contribute to, or detract from, scientific success. Although some of these may be considered products in-and-of themselves, such as peer review and scientific consensus, we are concerned with what features of tangles make scientific products reliable for what they are supposed to do, not with what social conditions, habits of mind, standards of conduct, and so on encourage the production of such features. Our contributions are also complementary to work on how scientific knowledge production is organised and what sustains or inhibits it, from the early studies in sociology by, for example, Ludwik Fleck ([1935] 1979) and Joseph Ben-David and Randall Collins (1966) on the role of communities in science to the recent work by Sabina Leonelli and Rachel Ankeny (2015) on repertoires—the 'assemblages of knowledge, social structures, methods, and tools', that is, the 'material and social conditions' (701) that are responsible—or not—for the resilience of scientific collaborations. All of this is essential if you are to understand and to monitor science, its procedures and its products. The right kinds of social and normative structures must be in place and function properly if science is to create virtuous tangles and thereby reliable products. We do not here attempt to add to the already excellent studies available of this. We tackle a different sort of job: unpacking the idea of a virtuous tangle, identifying what features characterise it, and explaining how those features make for more scientific reliability.

Our work has its beginning in the anti-theory turn that philosophy of science took in the late 1990s. The important deliverances of science are not all in its representations of the world—its theories, principles, and models. These representations by themselves cannot do much for us. As Nancy Cartwright (1999) argues, our scientific representations are not, as many hoped in the heyday of logical empiricism, a super vending machine to which we can feed in facts about the questions we are interested in, turn the crank of logical derivation, and pick up answers that fall out. Much, much more is required. But what?

The philosophy of science community went in two directions in answer.

One looks at other scientific products beyond representations. This began with models and what they do besides just representing. Concern with models in the second half of the twentieth century was stirred up by Mary Hesse ([1963] 1966) but then lay mostly dormant until the 1990s when new interest was spawned, in large part by the decades-long modelling project at the London School of Economics that helped generate the classic collection *Models as Mediators* by Mary Morgan and Margaret Morrison (1999). There the emphasis is close to ours, not primarily on the truth of models (or their truth likeness or isomorphism or similarity to the world) but rather on the purposes a scientific model is meant to serve and on whether it can be relied on to serve them.

Next, philosophers in this tradition turned to experiment, which, as Ian Hacking famously argues, 'has a life of its own' beyond just testing our representations. Then, to measurement, which we discuss in some detail in Section 1.5. And most recently, to data and data curation, spurred in part by work from Sabina Leonelli, whose roots are in the other post-theory tradition we are about to discuss.

The second tradition focuses on practice: what scientists actually do when they do science and in particular what they do to get science to succeed at the task at hand. This work has a particularist bent. The fine details matter to science's abilities to succeed, and philosophers of science should work to unearth and understand these details. There was concomitantly considerable stress on hands-on knowledge, know-how, and implicit knowledge—the kind of knowledge that is seldom articulated and generally not fully articulable.

This book builds on both these great traditions, extending and integrating them in a way that we hope provides a new take on a new formulation of the old question: 'what is it about science that makes it so good at what it does?'.

The focus on practice has directed attention to the details of the myriad activities that it takes to accomplish a scientific endeavour and how they work to do so. This kind of work is much needed. But we don't want the message to be lost in a focus on practices—where what is underlined is the fact that these are *activities*, things people do—because it misses out far too much of what we are concerned with. We use the term 'products' for the 'finished' components of science—finished in the sense of being accredited as reliable for the job on the label. These will include activities (e.g. the standard procedures for titration to determine the concentration of a substance in a sample that we are all familiar with from school chemistry). It includes these *and* ever so much more: models, measures, data sets, and the like are real *products* of science, and getting these to function properly is essential to scientific success and progress.

Our focus, then, is not only on the practices the sciences engage in but on all the products that science creates. But following the particularism of the philosophy-of-science-in-practice movement, we urge attention to the very great many, highly varied, highly refined products science deals with and to the fine details that distinguish one from another and that matter to their chances at reliability. We do so because, we argue, the reliability of any one of these in general depends not just on the fine details that circumscribe what it is but also on the reliability of a vast tangle of other products in the background.

When it comes to evaluating the *products* of science, we focus, as we have explained, on their *reliability*. This has not been much to the fore in either tradition. We do so in order to underline the importance of *purposes*. *Reliability* is a two-place relation: product P is reliable to achieve aim A. One and the same product can be reliable to do a certain job for one aim but not for doing that same job for another aim. We shall, for instance, talk a lot about measures and raise questions like those of Julian Reiss ([2008] 2016): is the CPI (consumer price index) reliable as a way to measure inflation for the aim of setting interest rates for monetary policy? What about for the aim of adjusting US veteran's benefits to keep them stable year to year?

The closest thing we have found in the literature to our idea of the tangle is Hasok Chang's (2012) 'system of practices'. These too are crafted to achieve specific aims. Chang, who is one of the founders of the philosophy-of-science-in-practice movement, offers the *coherence* of practices as key to scientific successes. He has put mammoth effort into telling us what coherence is and how to identify when practices cohere so that they can work together. We have nothing to add on that score. We concentrate, as we note above, not just on practices but on all the products of science that are required to underwrite the reliability of any one of them. And we offer an account not of whether they can work together but rather on what characteristics this body of products should have if you are to expect that when they work together they can reliably achieve their aims.

We think that this focus on the 'finished' products of science and the aims or purposes for which they have been vouchsafed as reliable is an important missing ingredient in much of the work on scientific practice. It is certainly the case that much depends on the specific activities that go into a scientific endeavour and how they work together, or not. But these activities involve the use of a vast number of scientific products that have been produced by other sets of activities. And these products must be reliable for doing the jobs they need to in the new practices in which they get employed.

There is, we believe, an important message not to be undervalued in the claim that science works by creating general knowledge, knowledge that is repeatable and re-usable, outside the context of its production and independent of the other products it works with there. It is often hard to see how this is true in the very excellent detailed studies produced in the philosophy-of-science-in-practice movement. Of course, our focus here is not on knowledge in the sense of representations of nature. But we still take it to be essential to the scientific enterprise that it create products that can be certified reliable for given aims and put on the shelf in the general store of science to be taken down and used in a host of different enterprises where those aims need to be achieved if the enterprise is to succeed. Recall our remark that we use the word 'product' deliberately to get you to think of the kinds of things you find on shop shelves: 'These have been envisioned, developed, created, assembled, and tested by one conglomerate of actors and put on the shelf to be taken down for use by different actors with different ideas and practices for employing them in their own enterprises.'

So, for us, the demand that science produce general knowledge cashes out in terms of its creating reliable products, products that do the labelled job for the labelled purpose. That a measure, or a model, or a concept is reliable to do this job for this purpose does not mean just that it can do it here in this particular scientific endeavour where it was created in consort with the specific set of products it worked with here. Rather, it can be relied on in a broad variety of other endeavours where something is needed to do that job for that purpose.

All this must be done with due care of course, and science—and scientists—often get it wrong. The label and the instructions for use, including the specialist know-how it may take to follow those instructions, may only pick out a reliable product in a far more restricted range of settings than indicated. And you have to read the label carefully. You can't, for instance, just take a statistical technique off the shelf and use it willy-nilly.

You've got to make sure all the conditions for its valid use are satisfied. Often the label is not very explicit. A kind of shared understanding that one has to be trained into is presupposed. Also, it is generally taken for granted that you may well have to make a lot of adjustments—corrections, addition, deletions—to fit the product to a new enterprise.[4] In Section 3.4.1 and following, we explain how important it is in doing this to figure out what we, borrowing from Hasok Chang, call the 'external purposes' of the enterprise. And often the labels are mistaken. The categories for which the product can be reliably used are specified too broadly, too vaguely, or using concepts that don't get real purchase on the world. Discovering and fixing this is all part of the normal everyday work of science. Finally, we should note that sometimes it is not altogether clear whether a product is on the shelf or not since scientists often disagree about the reliability of a product, sometimes even with good reason to back up both sides. We give an example of this in our 'Parting Thoughts' (Section A.6). In that case, clearly it is best to use the product with extreme caution.

Besides offering a new focus for both these two post-theory traditions in philosophy of science, we also aim to contribute to the scientific realism debate, which has its feet firmly in conventional concerns with theories, laws, and models, and their claims to truth. We hope that what we write will induce a redirection of energy from these representational devices in science to all the other marvellous products we discuss here. What virtues do we expect of them and why? And what secures those virtues? This book is about one absolutely central virtue that ensures their usefulness across different scientific endeavours—reliability, the ability to repeatedly do what it says on the label. We provide here one book-length account of what can help secure that virtue. But surely much more can be said if only philosophers direct their energies this way.

The overall point of Part I of this book is to show that the usual suspects, even in combination, fail to account for the amount of reliability found in the multitudinous products of science. Part II aims to argue that a tangle of background support is crucial and, as best we can, to characterise the tangle and to show how it functions to secure reliability. But we think we have a lot of other new ideas to offer on the way that we hope you will find worthwhile on their own. As we go along we will try to introduce some sign-posting so you can look at the trees in detail to see what they have to offer and nevertheless still see the forest grow.

---

[4] For nice examples of the kinds of adjustments that can be needed even for products that are very well-established—deemed as 'facts'—see Howlett and Morgan (2010).

PART I
THE USUAL SUSPECTS

1
Scientific Method

Preview

What follows is not a conventional discussion of scientific method. There is a great deal to be learned from the canonical literature on this topic, but there are excellent discussions and reviews of it, and anthologies already available (see e.g. Gower 1997; Gimbel 2011; Andersen and Hepburn 2015; Strevens 2020). There's no need for us to repeat those here. We'll get straight to the point. We don't like the idea of scientific method as usually conceived. That's for three reasons that we explain and defend in this chapter.

Objection (1):   Any characterisation that is broad enough to cover everything that reasonably gets counted as science, from gravitational wave theory to thermometry to Covid-19 vaccine research to measures of democracy to the theory of gender performativity and so much more, will be so general as to lack content. We discuss this relatively familiar point in Section 1.1.

Objection (2):   Using the term 'scientific method' is dangerous. Privileging methods narrows the scope of what science is up to and of what is needed to evaluate it. This is the topic of Section 1.3.

Objection (3):   Scientific method is too bound up with the attempt to establish scientific knowledge—true claims and explanations. But that leaves out the bulk of what science produces. For that the focus should be on reliability, and that requires a whole new research orientation in methodology and philosophy of science to study what supports and what undermines the reliability of *all the products* that science creates. This is key to our approach in this book. We defend our call for reliability in Section 1.4.

In 2017, Nancy Cartwright and Elliott Sober won the Lebowitz Prize from the American Philosophical Association and the honorary academic society Phi Beta Kappa for their debate 'Is there such a thing as the scientific method?'. This debate will play a central role in both Objection (1) and Objection (2). The lectures for the debate are supposed to offer 'contrasting

The Tangle of Science: Reliability Beyond Method, Rigour, and Objectivity. Nancy Cartwright, Jeremy Hardie, Eleonora Montuschi, Matthew Soleiman, and Ann C. Thresher, Oxford University Press. © Nancy Cartwright, Jeremy Hardie, Eleonora Montuschi, Matthew Soleiman, and Ann C. Thresher 2022. DOI: 10.1093/oso/9780198866343.003.0002

(not necessarily opposing) views…The dialogic character of…[these lectures] emphasizes the historic work of philosophy as a process of inquiry'.[1] In the case of Cartwright and Sober, the dialogic character of the discussion achieved its goal. The process of enquiry resulted not in defences of opposing views but in refinements of the question and of the answers. Sober's take on scientific method in this debate figures prominently in our first objection, Cartwright's in our second. This debate is not published anywhere so we describe it in some detail to make it available to readers.[2]

We begin with some general remarks about the traditional idea of scientific method that lay the groundwork for our three objections, especially Objection (1) and Objection (2).

1.1 Where's the Bite in the Scientific Method?

'Scientific method' is commonly used in three senses:

(1) A *procedure*—a general strategy to be followed in order to achieve some scientific goal (a discovery, an explanation, a sound measure, generally the solution of some scientific problem).

(2) A *set of rules or principles*—prescriptions or norms that discipline the following of the procedure.

(3) A *set of techniques*—conceptual or practical tools that implement the procedure.

In the sense of (1) a method is a course of action, or a way of reasoning, in view of an aim. The usual take on scientific method—with which we take issue in this book—supposes that the general aim of science is objective knowledge, or truth, and the way to get there includes one or more steps.

Here is an example. Suppose you want to know what sunspots are. Galileo advised: perform systematic observations, put forward a conjecture, draw logical consequences from the conjecture, test the consequences by means of new observations, reject conjectures that imply false observation, and accept the one that remains after many different observations of the right kind (Galilei [1632] 2001, Third Day).[3] This has widely become known as the

---

[1] See https://www.apaonline.org/page/lebowitz.

[2] We are grateful to Sober and Cartwright for making material available to us to reconstruct the debate.

[3] For a discussion of Galileo's method, as well as the description of the three meanings of 'method', see ch. 1 of Pera (1994).

Galilean method, which is followed in many domains of scientific research. It can be considered a good candidate for a scientific procedure.[4] We note for future reference that the fifth step—reject conjectures that imply false implications—was famously endorsed by Karl Popper ([1935] 1963) under the title 'falsificationism'.

Why do humans have a backbone while 90 per cent of animals don't? Here is how you might proceed to answer the question: look at the available evidence, formulate a number of explanations, and infer from the available data the explanation that best accounts for it. This procedure goes under the label *inference to the best explanation* (IBE). How to prove a mathematical theorem? Draw consequences from a set of axioms, which is the *deductive method*. How to figure out if the kink in the APRES data supports the hypothesis that high temperature superconductivity is due to phonons? The *hypothetico-deductive method* (HD) tells you to see if the kink is a deductive consequence of the hypothesis when the hypothesis is supplemented with acceptable auxiliary hypotheses. How to cure hysteria? Freud says: release repressed emotions and experience, reveal traumas and unresolved issues, make the unconscious conscious (*psychoanalytic method*). Or, looking forward to our discussion in Section 2.3, how can the effectiveness of a policy in a target population be measured? Here is one widely adopted procedure: evaluate the difference between what the average effect size would be if the population were subjected to the policy versus what the average effect size would be if it were not. This is called *the counterfactual method*.

However, the general formats just offered do not say anything about how the procedure is to be carried out. This is what (2) is meant to secure. Rules and principles provide the procedure with a code or set of guidelines for how it might be implemented—they tell you how to carry out each step in the Galilean method, how to falsify, how to determine the best explanation, how to judge auxiliaries acceptable in drawing consequences from a hypothesis, how to make the unconscious conscious, or how to apply the counterfactual method to find an average population effect size.

Consider the fifth step in the Galilean method. How can falsification be implemented? Causal-process-tracing methodology tells you to pick cases to

---

[4] Sometimes in the broader literature on science the Galilean method is also called the 'hypothetico-deductive method' (HD), and has been interpreted in that way by a number of Galilean scholars in the huge philosophical literature on Galileo's science. See, for instance, the articles by McTighe and Settle, both in McMullin (1967); also Drake (1972, 1973); Shapere (1974). In philosophy, HD has come to be most commonly used to label a model of explanation (a hypothesis explains phenomena that can be deduced from it) and next most commonly as we use it here and in Section 4.1, for a method delimiting the kinds of empirical facts that can be used in support of a hypothesis (those that follow deductively from it).

study where the hypothesis is most likely to fail if false. Also 'Be relentless in gathering diverse and relevant evidence' (Bennett and Checkel 2015, 27).

Next, looking at the case of inference to the best explanation, by what criteria can 'best' be defined? There are a number of principles commonly offered in answer and they do not always yield the same result. One way is to choose the simplest (sometimes called the most 'parsimonious') explanation; another is to choose the likeliest in probability terms; yet another is to choose the explanation with the most unifying power, and so on.

In the case of the HD method, by what criteria are auxiliaries to be judged acceptable? Here there are two principles that are standardly appealed to: 'Auxiliaries should ideally follow from, or at least be visibly consistent with, high level theory' and 'Auxiliaries should be empirically well confirmed.' These too, like those for IBE, do not always point in the same direction, as Di Bucchianico argues was the case for the warring camps in superconductivity introduced in the Preface.

How to make the unconscious conscious in psychoanalytic method? Freudian rules of analytic interpretation guide the therapist here.

How can you evaluate the average counterfactual difference in a population between receiving a treatment and not receiving it? As we explain in Sections 2.3 and 2.4, the central principle underwriting how to implement this is in the form of a theorem: if all other causes are orthogonal to the treatment, the difference in average outcome between a sample of the population that receives the treatment and one that doesn't is an unbiased estimate of the counterfactual difference.

If you now look at (3), method is here pictured as techniques or tools. How do you pick the cases to study that are most likely to fail if the hypothesis is false? Standard advice says to pick cases that the hypothesis should cover but are very unlike those that suggest it in the first place.

With respect to parsimony, once it is decided that it is the simplest explanation that one is after, how can simplicity be pinned down? That depends on what exactly is being studied and on what concept of simplicity is taken to be appropriate to it. You could do so by using some technique for, say, counting causes, as in the backbone example Sober discusses, or in case the explanatory hypothesis is formulated in terms of an equation, by counting the parameters, or for formalised theories, by seeing how many axioms are postulated.

How can auxiliaries be empirically established? Techniques to be used here again depend on the content of the auxiliary. Causal claims require different techniques than correlational, psychology experiments have different protocols from experiments in physics, confirming the value of a fundamental constant requires yet a different kind of technique, and again, what that looks like will be heavily content dependent.

What do the rules of analytic interpretation dictate for making the unconscious conscious for this patient? Perhaps the technique of free association or dream interpretation.

How can one cause be made orthogonal to other causes in two sample populations? One standard answer, as we describe in Section 2.3 in discussing the methods of the randomised controlled trial (RCT) is by random assignment to the two groups and careful policing after for confounding.

Why organise 'the scientific method' in these layers—procedures, rules/principles, and techniques—and what can be learned from our discussion of them?

Science is supposed to serve as a model of good reasoning. In order to do so, 'the scientific method' should display sufficient prescriptive content to make its dictates adequate not only as rules of good reasoning but also as rules of good reasoning in the specific circumstances of a specific subject matter of a specific scientific research agenda. For example, if you use the rule of parsimony to choose a good hypothesis, the rule should tell you not only how to choose hypotheses 'in principle', but also how to choose this particular hypothesis with these particular features in this particular context with this particular aim in mind.

To achieve adequacy in this sense, the scientific method should specify what is to be done not only in general but also in concrete circumstances. For instance, how are you to decide if the hypothesis of common ancestry counts as the simplest explanation in the case of the human backbone? But by adding precision to meet the requirements of specific contexts, the universal prescriptive force decreases. In the end, your prescriptions might be made adequate to one situation but not to another. They thus cease to be prescriptive in the sense that general rules and principles are required to be.

This is what happened with Galileo. In the Second Day of the *Dialogue* Galileo claims exactly what Popper puts at the core of his falsificationist method: hypotheses that logically admit of one contrary observation should be ruled out.

For I understand very well that one single experiment or conclusive proof to the contrary would suffice to overthrow both these and many other probable arguments.   (Galilei [1632] 2001, 122)

Still, what type of observation is able, on its own, to knock down a whole theory, one might want to know? Elsewhere (*Letter to Francesco Ingoli* 1624)[5] Galileo further refines the rule by saying that we ought to rely on 'established' experience. But could any single established experience, or collection of them, play this falsificationist role? In the same letter Galileo specifies that only an experience that does not count as a 'local or secondary anomaly' can play a falsificationist role. Galileo is prepared to treat as local and secondary some of the objections to the Copernican theory (the very same objections that Ingoli by contrast considered crucial and final) by looking at them via the eye of further assumptions and ad hoc hypotheses that help him save this favourite theory of his.

So it looks as if Galileo attempts to make the rejection rule more and more precise, but by making the rule adequate to the context of interest, he also makes the rule lose its universal force. He does not hesitate to adopt a principle of tolerance that saves the rule in context but gives up on its prescriptive force across contexts.

Similar difficulties in sticking to general applicability can be raised for principles, for procedures, and for techniques.

As to principles, should you expect them to hold generally, everywhere the procedure that they inform applies? Not usually. In many cases this is obvious since the same procedure can be accompanied by oppositely pointing principles. Think of the procedure of choosing the hypothesis that provides the best explanation. The principles that help constitute what is 'best' are hotly contested. Parsimony is one, but 'look for heterogeneity' is also championed, both by many feminists and by those who think of nature as fundamentally complex and diverse. Or consider Lorraine Daston and Peter Galison's history of objectivity (Daston and Galison 2007). In the era they call 'truth to nature' the ideal was parsimony: look for the one true form that underlies the superficial diversity you observe around you. But the next era, they note, which they dub the era of 'mechanical objectivity', aimed to represent nature in all of its heterogeneous detail.

Looking at practices across the sciences it seems that neither one choice nor the other is right. In deciding what is the best explanation in a given context for a given phenomenon, sometimes the principle of parsimony is invoked, sometimes heterogeneity. As we see Sober arguing in Section 1.2, parsimony may be a candidate for applicability across the sciences, but whether it should be applied in any given context will depend on context-specific information.

---

[5] Partially translated in Graney (2015).

As to procedures. First, the choice depends on the aim of the enterprise. IBE is in aid of choosing a hypothesis, as is the general Galilean method; HD is about determining what facts can count as evidence for a hypothesis; the psychoanalytic method is in aid of relieving psychological distress. Even considering a single aim, there is still choice on offer. For instance, with respect to choosing a hypothesis, IBE is far more liberal than falsification-ism, at least as understood by Popper. Popper never allowed acceptance of a hypothesis unless it is the last one standing—which one can never be sure of since there may always be more hypotheses that we've not yet thought of.[6] IBE can also yield different results from both of the methods of hypothesis confirmation we discuss in Section 4.1—'HD-with-bells-on' and various probabilistic procedures. Given that there are options, making one choice and demanding it generally across the sciences and across subject matters does not seem the best strategy. Better to allow content-dependent case-by-case choice.

Even more so in the case of techniques, generalisability is neither a requirement nor a necessity. Techniques can be specific to a domain of research (e.g. free association, or applicable to more than one domain); for example, many of those that are thought of as 'quantitative' in the social sciences, like game-theory modelling and statistical techniques (e.g. the ones Sober advocates) or those that get classified as 'qualitative', like interview, questionnaire, and focus-group methods. In either case, as we argue follow-ing Sober, their applicability depends on the context of use[7] and on the aims of the procedures they are used for.

We draw a number of lessons from our discussion. The first is about uni-versality. It is in support of our Objection (1) to the whole idea of 'the scien-tific method'. If you insist that what constitutes the scientific method fits all sorts of scientific contexts, 'the scientific method' inevitably becomes general to the point of being vague, incomplete, and ultimately uninformative and unusable.

This is in no way a new point. It is, for instance, defended in great detail by Hilary Putnam (1981), where he concludes that any characterisation broad enough to get in everything one might reasonably count as science would end up including other areas where critical discourse matters, like ethics.

In the Lebowitz debate that we discuss next, Sober makes the same point, coming at it from a different angle. Is there such a thing as the scientific method? Sober answers, yes and no: the glass is half full *but also* the glass is

---

[6] As Stanford (2006) brings to the fore on unconceived alternatives.
[7] According to criteria of use that no doubt constitute a specific problem for scientific inquiry.

half empty. There are some reliable methods that can be used widely across the sciences. But whether they can be used in a particular case and if so, just how to use them, always requires subject-specific input. We see this as a general feature of representations anywhere that there is a great deal of phenomenological diversity, whether the representations are of the world itself or of the vast array of different sciences with their vastly different methods and practices that we use to study the world. To say something that is reasonably broad, covering a number of diverse-looking cases, you need fairly abstract language, concepts whose concretisations look different in different cases. But then the very representation that holds broadly is not enough by itself to tell you what that amounts to in the concrete in any of the cases it covers. This is our diagnosis of what is happening in Sober's examples. When methods are characterised in a general enough way to cross domains, the descriptions become so general and abstract that they don't tell you what you need to do. The specific techniques needed are missing.

Sober's point is also one that Putnam makes in different terms in criticising what he calls 'method fetishism': it is not possible to separate the content of science from its method. The method changes (it is not unique) depending on changes in content. Even the strongest exemplars of formally designed methods—falsification and Bayesianism—give different answers depending on the concrete assumptions on which they rest. For example, the results of applying some form or other of Bayesianism depend on how the prior probabilities are established in any given context, and this is something that the formal design itself cannot provide, as we note in Section 4.1 (Putnam 1981, 190–1).

Our *second lesson* is about the almost universal impossibility of ever saying enough (a point we return to in Sections 3.3.2 and 3.3.4). You will have noticed that our examples of techniques are still not very specific. The techniques for implementing an RCT tell you to assign subjects randomly to two groups. How do you randomise? By flipping a 'fair' coin? How do you ensure the coin is fair and the flips are done in the right way? By a random number generator? Which? Bought where? And then what do you do specifically to get the assignments dictated by the random process actually made? You are supposed to be achieving 'orthogonality' to other causes. But there are many things that can go wrong, sometimes even deliberately. For instance, in a section headed 'Simple methods of randomization (and how to subvert them)', Andrew Vickers (2006) reports:

> This is not merely a theoretical problem: Ken Schulz has documented various ways in which clinical researchers have attempted to subvert randomization, including bright light sources to reveal allocations in sealed envelopes and even

ransacking the principal investigator's office to find the randomization list... My own favorite anecdote illustrates another aspect of randomization that must be protected. In a trial of social support for women with at-risk pregnancies, randomization was implemented by having a researcher pick a marble from an urn, blue for treatment, white for control; in practice, however, if the research nurse picked out a white marble for a patient she felt really needed social support she simply replaced it and selected another one. As a result, baseline distress scores were higher in women in the treatment group: this would make the trial more likely to find [*sic*] a difference between groups, even if none existed, due to regression to the mean.   (195)

Two things stand in the way of getting totally specific. First, the right way to do it in one case won't be the right way in another. And second, as we stress in Sections 3.3.2 and 3.3.4, it is almost always possible to 'game the system'— to do exactly what is dictated and still not act in the spirit of the ends those dictates are meant to serve. You'll see that a closely related problem arises with respect to specifying the aims to which science products are put when we discuss 'The Monkey's Paw' in Section 4.5.

Our remaining lessons work together as a triumvirate. The *third lesson* is that choices are required all the way up and down the layers. Inference to the best explanation among a set of hypotheses may not be a good idea if you don't have good reason to think that you can pick criteria for 'best' that track the truth. Perhaps it is better, then, just to attempt Popperian falsification for each of the hypotheses. Drawing balls from an urn may not be the best way to randomise if you can't ensure that none are surreptitiously slipped back for a redraw. The *fourth lesson* is that good practice should require that choices be justifiable. In Section 1.2 you see Sober showing how this might be done for the principle of parsimony. The *fifth lesson* is that the justification will require context-specific information, which is at the heart of Sober's claims.

We turn now to look at Sober's discussion in detail.

1.2 Dealing with Objection (1): What's Really There in the Empty Half?

Sober illustrates his position using the principle of parsimony.[8] Parsimony is worth spending a little time on because it is a classic example of what

---

[8] The view supported by Sober in his dialogue with Cartwright was inspired by Sober (2015), written for a wider audience.

philosophers of science call an 'epistemic' virtue. Others include breadth of scope, explanatory power, unifying power, variety of evidence, and novel prediction. These are all taken to be part and parcel of the methodological principles used to settle on the truth or acceptability of theories and models across the sciences.

The principle of parsimony has a long and well-established history, which speaks in favour of its general applicability and of its normative force. Already Aristotle in *Posterior Analytics* attributes superiority to demonstrations that derive from fewer postulates or hypotheses.[9] He also believes in parsimony as an ontological principle. For instance, he rejects Plato's theory of forms by making use of an argument centred on parsimony. But the name most associated with parsimony is no doubt the Medieval theologian, philosopher, and logician, William of Ockham. Ockham stated the principle in different forms,[10] but he mainly conceived of it as an epistemological principle that allows you to choose the best theory (known as Ockham's razor). Ockham himself used it to reject Buridan's impetus theory of motion (Crombie 1953, 176).

Newton famously stated the principle in his first rule of philosophical reasoning in his *Principia* (Bk III): 'Rule I: We are to admit no more causes of natural things than such as are both true and sufficient to explain their appearances' (Newton [1687] 1999). Newton also expresses the rule in decisive ontological terms: 'Nature is pleased with simplicity, and affects not the pomp of superfluous causes.' Similar echoes of the idea of parsimony are to be found in Newton's other three rules—the third and fourth (about experiments and about induction) rest on the ontological idea that nature 'is wont to be simple', while the second appeals to a more epistemological view concerning causal theories: 'To the same natural effects we must, as far as possible, assign the same causes.'

Ockham's razor was also famously endorsed by Kant, who in the *First Critique* referred to it as a regulative idea of pure reason widely adopted by scientists. In more recent times it was also appealed to by Einstein. For him, physics should aim at explaining the widest number of facts from the

---

[9] 'We may assume the superiority *ceteris paribus* [other things being equal] of the demonstration which derives from fewer postulates or hypotheses': Aristotle, *Posterior Analytics*, in McKeon (trans.) (1941), 194, 150.

[10] For example, Philotheus Boehner, a noted Ockham scholar, wrote of Ockham's razor: 'It is quite often stated by Ockham in the form: "Plurality is not to be posited without necessity" (*Pluralitas non est ponenda sine necessitate*), and also, though seldom: "What can be explained by the assumption of fewer things is vainly explained by the assumption of more things" (*Frustra fit per plura quod potest fieri per pauciora*). The form usually given, "Entities must not be multiplied without necessity" (*Entia non sunt multiplicanda sine necessitate*), does not seem to have been used by Ockham' (Boehner 1964, xxi).

smallest number of assumptions and axioms.[11] Indeed, it was this principle that he claimed led him to the discovery of his general theory of relativity. Equally, it is steadily invoked in contemporary philosophy as a central guiding principle across the sciences for choosing theories and models.

In the history of science one can find different versions of the principle being used widely but with mixed success. Copernicus famously justified his choice of a heliocentric theory on grounds of parsimony. Both heliocentric and geocentric theories 'fit the data', but the heliocentric theory does so with many fewer epicycles and in a form far more harmonious and elegant than the geocentric (Figure 1.1). Galileo made a less successful use of the

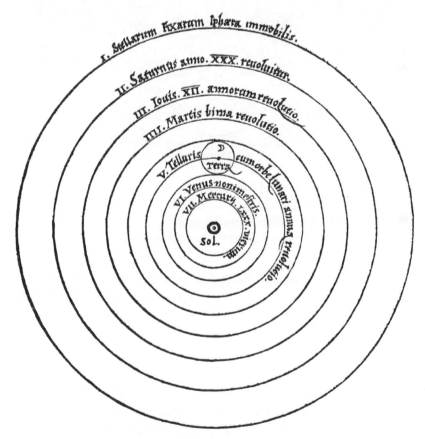

Figure 1.1 The heliocentric model of the solar system, drawn by Nicolai Copernicus in *De Revolutionibus Orbium Coelestium* (1543)

Source: https://commons.wikimedia.org/wiki/File:Copernican_heliocentrism_theory_diagram.svg.

[11] 'It can scarcely be denied that the supreme goal of all theory is to make the irreducible basic elements as simple and as few as possible without having to surrender the adequate representation of a single datum of experience' (Einstein 1933, 165).

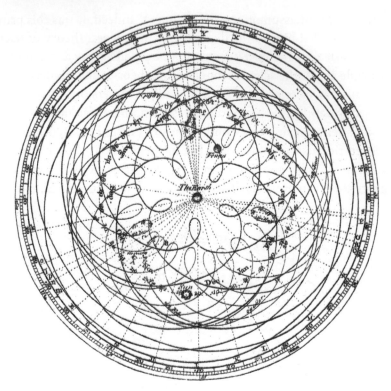

Figure 1.2 The apparent orbits of the sun and planets on a geocentric model from
Encyclopaedia Britannica (1st edition, 1771; facsimile reprint 1971), Volume 1, Figure 2
of Plate XL facing page 449. Artist: James Ferguson
Source: https://commons.wikimedia.org/wiki/File:Cassini_apparent.jpg.

principle when he postulated that the concept of 'circular inertia' is the only
type of motion, arguing on the basis of parsimony (Figure 1.2). For him rec-
tilinear motion was just an illusion, a claim that was ultimately rejected
(Gauch and Gauch 2003, 274).

Parsimony is, then, firmly at the centre of the description of scientific
method within well-established philosophical and scientific traditions. Is parsimony a reliable principle of scientific method, as Newton and others recommend? Is it generally applicable? Let us see what Sober has to say about this.

For Sober, science can employ general frameworks—ones that apply
broadly across many areas of science—for determining whether and why
parsimony is relevant. But then, within each framework, the justification for
using Ockham's razor depends on specific empirical assumptions that may
be true of the particular problem addressed and not of others. Sober illustrates with the distinction between common-cause and separate-causes
explanations in ancestry. If both humans and monkeys have tailbones,

assuming a common ancestry (CA) hypothesis is more parsimonious than admitting a separate ancestry (SA) one. But why is parsimony relevant in evaluating the two hypotheses?

Another general principle, used broadly across many sciences, comes to the rescue, offered by Bayesian theory. What is the probability of an observation O on the assumption that a certain hypothesis H1 is true versus the probability of the same observation on the assumption that an alternative hypothesis H2 is true?

- O favours H1 over H2 if the Prob (O/H1) > Prob (O/H2)

So here is how we reason, according to Sober:

A. If Prob (Humans and monkeys have tail bones/CA) > Prob (Humans and monkeys have tail bones/SA)
B. Then CA is more probable than SA.

But how can science decide which of these two conditional probabilities is greater? To do so, science relies on specific assumptions about the subject matter at hand, says Sober. So, what assumptions justify the conclusion that the probability given CA is higher than the probability given SA? Sober suggests the following analogy.

To decide whether Sue and Sally are sisters or whether they are unrelated you could rely on the observation that they both have gene G. This might appear as evidence in favour of common cause. But suppose that gene G renders individuals sterile after their first child. This is evidence that Sue and Sally are not siblings, and therefore that it is more likely that each having G has separate causes rather than a common cause. So, Sober concludes, though a common cause is a more parsimonious hypothesis than separate causes, whether it is more likely depends on particular subject-specific assumptions. This results in a 'conditional justification' of adopting Ockham's razor as a methodological principle.

Sober sees a general pattern emerging. The principle of parsimony is widely applicable. It is used, and used successfully, in almost every domain of science. But it does not apply in every case. Whether applying Ockham's razor is justified in any particular case depends on assumptions about that specific case. This is not an argument against the general applicability of the principle of parsimony. It is an argument in favour of the applicability of the general principle in particular cases.

Sober is inclined to believe that what can be argued for parsimony and likelihood reasoning holds for most methodological principles that apply across disciplines, time, and subject-matters. They all need *substance specific information*, first to *justify* applying them to a specific problem, and second to determine *how* to apply them to that problem. The glass is indeed half full because there are methods that can be used broadly across different sciences. But it is half empty because it requires these two kinds of subject-specific information to apply them.

How do you vouchsafe this usually very concrete subject-specific information? Presumably each piece of information and each premise used in the justification that the general method applies in this case must itself be secured—by reliable methods. So Sober's glass is half full because there are a number of methods, like parsimony, that can be used broadly across many sciences. And we are happy to suppose for the sake of argument that there are good reasons to think these methods are reliable for doing what they are supposed to in any given case if they apply to that case and are properly used there. But Sober's half-empty glass is, it turns out, really a Pandora's box that spills out a multiplying proliferation of more and more methods, calling for more and more defences that each is up to the job it is set to do in each specific case.

Recall why we are discussing scientific method in the first place. Our issue is 'What makes science so reliable?'. One contributing factor is supposed to be that science uses either *the* scientific method, which is supposed to have much to say in its defence, or it uses *reliable* methods. We have no quarrel with this. We think the use of the scientific method, whatever that is—or of reliable scientific methods—is indeed part of what makes science products reliable, just as we agree that the other two usual suspects also contribute greatly to science's successes.

But you can't stop there. What makes these methods reliable? Maybe they are defended by premises that are themselves vouchsafed by reliable methods. That we hope is true. But that cannot be the end of it. What ensures that the concepts employed in the subject-specific information needed and the concepts employed in the defence of the methods make sense, that the measures that are employed measure what they are supposed to, that any approximations used are appropriate, that data collection and curation are up to the job in the case at hand, and so on?

The answer to this, we argue, catapults you right into the need for what we call 'the tangle of science'. The fitness for the role that Sober's universal principles need to play in any specific case is secured by a vast network of other scientific work that defines the concepts used in them, develops measures for them, assesses the validity of these measures, tests the claims implied once the concepts are well defined, and much more. That is, the reliability of the usual highly favoured methodological principles to guide science to true hypotheses and correct explanations depends on each occasion of use on just the kind of tangle we introduce and describe in this book.

1.3 Dealing with Objection (2): Privilege Is Pernicious

Parsimony is, along with fruitfulness, unifying power, novel prediction, internal coherence and many more, in a set of what philosophers call 'theoretical' or 'epistemic' virtues.[12] 'Epistemic' means knowledge conducive—a guide to true hypotheses, correct explanations, descriptively accurate models. Whether any or all of these are epistemic or not is notoriously a source of controversy. We take a view in this book on epistemic virtues in general very much like Sober's on parsimony. Whether one or another of these is a guide to truth on any particular question or occasion depends on what the facts are. Often, for instance, the most fruitful models are not true to the facts but involve strong idealisations.[13]

Epistemology aside, are these features you should want in science? That depends on what you are trying to do. Any one of them may serve some purposes well and others not at all. Regardless, they should not carry the honorific 'virtue'—or so argues Cartwright in the debate with Sober with ideas that we build on here. Her worries are one of the reasons we think that, despite the importance of having reliable scientific methods, talk of 'the scientific method' should be avoided.

As we noted, we do not in any way disagree with Sober that there are a number of methods in science that can be used broadly across scientific disciplines and subject matters. And that is all to the good. It allows science to hone the methods and learn general lessons about their use. What we resist is a slippage from 'Here are some methods that are universal (or at least widely applicable) in science' to 'These methods help characterise science' to 'These are methods you should make effort to use if you want to do proper science.' We want to stress that the fact that a method is universal does not give it any special status. Wide applicability may mean that the sciences are more practised in its use but it does not mean that it is a better method to use. It may seem a truism but it is one worth underlining that what are the best methods to use depends on what the problem is, on what knowledge is available or can be reasonably obtained, and on what skills and resources are to hand.

---

[12] As first discussed by Kuhn (1973) and then further discussed, among others, by Laudan (1984); Longino (1990); Keas (2018); Schindler (2018); Douglas (2000); Bhakthavatsalam and Cartwright (2017).
[13] See, for example, Batterman (2009); Elgin (2009). Of course, some, like logical consistency, may be universally necessary for truth.

This echoes Cartwright's concerns in the Lebowitz debate, where she warns about the dangers of too much reliance on any particular methods. Cartwright objects to identifying any as *the* methods to use in a given domain or to use for a given kind of problem and certainly none should be taken as definitive of science itself: no methods should ever be thus privileged.

There are two reasons, both of which we endorse and develop in later sections. First, the methods that science uses fix the kinds of features that it studies and the kinds of questions it can answer. Science should *not legislate* but rather *discover* what there is and what questions to ask. Second, the kinds of methods that promise highly reliable outputs, given reliable inputs, generally answer very narrow questions. These are the methods we label 'rigorous'. They can do very little for us precisely because they are narrow, as we argue in Section 2.4.

Not only does choice of method dictate what questions can be asked but you can all too readily be tempted to slip from the question that can be answered with your favourite method to nearby ones that cannot. This happens often in the case of RCTs. We discuss these in detail in Section 2.3, which you can turn to for an account of what they are and why they can do what they can do.

What matters for the purposes of Cartwright's arguments about scientific method is that RCTs can answer questions about *causal ascription*: they can provide information about whether a given cause has had a given effect in the population studied at the time it was studied.[14] They cannot answer questions of *causal prediction*—will it have that effect in the same population at a different time or in some different population? Nor can they answer questions of *causal generalisation*—does it always or often have that effect? There are many good research designs for investigating these but RCTs play only a small role in any of them. Nevertheless, it is not unusual to find suggestions that RCTs can answer these other questions almost on their own.

Consider these remarks from Esther Duflo and Michael Kremer, who along with Abhijit Banerjee won the Nobel Prize in economics in 2019 for encouraging RCTs in development economics:

> The benefits of knowing which programs work...extend far beyond any program or agency, and credible impact evaluations...can offer reliable guidance to international organizations, governments, donors, and...NGO's beyond national borders.   (Duflo and Kremer 2003, 30)

---

[14] In the ideal and in the long run, as we show in Chapter 2.

Impact evaluations are to establish causal ascription—something RCTs can do. But 'which programs work' is causal generalisation and 'reliable guidance' is causal prediction, neither of which can be delivered by an RCT. Yet the three flow seamlessly into one another in just one sentence, as if they are the same question that can be answered with the same tools.

There's nothing special about RCTs though, nor about these scientists. To the person who has invested in a costly hammer, every job looks like pounding in a nail. Scientists are prey to this just like the rest of us. Privileging some methods over others encourages science to focus on jobs those methods can do, to try to use those methods for jobs they can't do, and to leave on the shelf—or worse, toss in the bin—methods that can do better.

Although our talk of scientists being 'encouraged' may sound psychological, we should note that this is a logical point, not a psychological one. A good precisely specified method can establish only results of a certain *kind*. So, if you privilege some set of methods—for instance claiming they are the best or most rigorous or for some reason the only ones to use in proper enquiry—you necessarily thereby circumscribe the kinds of results that can be established.

1.4 Dealing with Objection (3): From Truth to Reliability

Philosophy of science has been much taken up with issues of scientific method: just what is it that makes science so good at what it does? In answer, philosophers of science propose lists of epistemic or theoretical virtues like those mentioned in Section 1.1: novel prediction, variety of evidence, severe testing, exclusion of rival theories, and the like. They debate whether theories can be confirmed or, as Karl Popper argued, only falsified. Assuming confirmation is possible, they offer a host of accounts of how to infer from data to theory, like formal schemes of inductive inference, confirmation theories of various ilks, or inference to the best explanation. Philosophical accounts of scientific method are geared to philosophy's general focus on truth and theory, on knowledge claims and the ways to secure them.

It is important to bear in mind how seriously pervasive and consuming this focus is, so we'll spend a little time illustrating before proceeding with our main points. Much of what follows immediately will be old hat for philosophers of science, but other readers may find it helpful to get a sense of the kinds of issues that preoccupy philosophy of science and how centrally they focus on claims and theories.

Consider, for example, the huge amount of effort that has gone into articulating, defending, illustrating, refining, and disputing the famous argument in philosophy of science called the *pessimistic meta-induction*. This was first formulated by Larry Laudan (1981) and has been widely discussed in the literature either as a *reductio* or as an inductive argument. A classic *reductio* version is due to Peter Lewis (2001, 372):

1. Assume that the success of a theory is a reliable test for its truth.
2. Most current scientific theories are successful.
3. So most current scientific theories are true.
4. Then most past scientific theories are false, since they differ from current theories in significant ways.
5. Many of these false past theories were successful.
6. So the success of a theory is not a reliable test for its truth.

As a plainly inductive argument the pessimistic meta-induction looks like this (e.g. Wray 2013; Stanford 2006): many scientific theories in the past turned out to be false, despite appearing at the time as successful as present theories, so it is reasonable to expect that our own theories, despite current success, could also turn out to be false.

In either formulation, Laudan's argument emphasises that the move from the success of a theory to its truth is unwarranted. The historical record of a theory's success is no indication of the theory being true. The argument is powerful especially because it puts all defenders of successful theories on a par: defenders of theories in the past reasoned exactly as we do nowadays, so there is no reason to believe that we are on a better footing today vis-à-vis our own theories (Stanford 2006, 7).

Another huge topic in philosophy of science is the underdetermination of theories by data. This again is concerned with claims or theories and their truth or acceptability.

Here the underlying idea is that the evidence available at any given time in support of a theory is never sufficient to determine that the theory is true. This might depend, as Pierre Duhem ([1906] 1982) and then W.V.O. Quine (1951) pointed out, on the fact that theories never operate in isolation but always in conjunction with other hypotheses—what we have called auxiliary hypotheses—or background assumptions. So you never know whether a failed prediction results from the theory being false or from the failure of some of the background assumptions. (We have much more to say about these background assumptions in Chapter 4.) Underdetermination also might depend on the fact, regularly brought forward in the literature, that there are always alternative theories, whether already conceived or not, that are equally supported by the same evidence. Despite the impressive record of a theory in deriving predictions or explaining facts, none of these achievements can speak in favour of its truth over the theory's empirical equivalents.

Hundreds and hundreds of papers have been written in philosophy of science about how to think about truth and acceptability in the face of underdetermination and the pessimistic meta-induction. Meta-induction is, after all, induction, so its conclusion is never certain. It is a type of reasoning that has led people astray on many occasions—why should it score better here? Or, it can be pointed out, scientific theories in the past might indeed share some of the virtues of present theories but to a lesser degree or under a different conception of success, so theories are unfairly evaluated under a unified blanket that misses altogether significant differences among theories in different periods of time (Stanford 2006, 10).

Just as much attention goes to underdetermination. First, it is pointed out that for at least some theories it is difficult to warrant that there are actual alternatives that are not trivial or even possible alternatives that are not trivial or highly implausible given what we know of nature, so in the absence of such warrant, underdetermination is just a highly speculative possibility (Kitcher 1993; Leplin 1997). It might be conceded that there are some convincing examples where there are such alternatives. But these all come from the physical sciences (typically, the relativity of motion, as pointed out in Laudan and Leplin 1991), and it might be objected that this is not a sufficiently or significantly representative pool of theories from which to extrapolate a general thesis. Second, in the absence of such warrant, lots of scholarly effort has been put into producing procedures that can generate alternatives 'in principle', in the form, for example, of algorithms for formulating empirical equivalents to any hypotheses (e.g. Kukla 1996). However, turning the issue into an 'in principle' problem does not settle whether alternatives produced this way have any bearing on how theories are chosen in practice, or how they should be.

Further recent discussions draw attention to 'unconceived alternatives' (Stanford 2006; Sklar 1981). Why suppose that the best among the alternatives we know is likely to be true or near true, even supposing it has proven highly successful, when we know there are so many alternatives that we have not yet conceived, and many we never will?

This is just a cursory overview of how philosophers of science put theory, hypotheses, truth, knowledge, and acceptability at the heart of their concerns.

Nor is this focus on theory and knowledge, and getting those right, peculiar to philosophy of science. For instance, the US National Academy of

Sciences' 2008 report *Science, Evolution, and Creationism* builds this right into their *Definition of Science*:

> The use of evidence to construct testable explanations and predictions of natural phenomena, as well as the knowledge generated through this process.[15]

They go on to describe 'how science works':

> Scientific knowledge and understanding accumulate from the interplay of observation and explanation. Scientists gather information by observing the natural world and conducting experiments. They then propose how the systems being studied behave in general, basing their explanations on the data provided through their experiments and other observations. They test their explanations by conducting additional observations and experiments under different conditions. Other scientists confirm the observations independently and carry out additional studies that may lead to more sophisticated explanations and predictions about future observations and experiments. In these ways, scientists continually arrive at more accurate and more comprehensive explanations of particular aspects of nature.

Similarly, the UK Science Council claims:

> Science is the pursuit and application of knowledge and understanding of the natural and social world following a systematic methodology based on evidence.[16]

So—it seems—science is all about knowledge and explanation. And accounts of scientific method aim to explain why science is successful in finding these.

We have no quarrel with propositional knowledge and explanation, and we agree that much of what is mentioned above can sometimes be conducive to achieving them. But we lament the extreme narrowness of focus.[17] Theories, knowledge claims, and explanations are only one small item among the panoply of things that science produces, that serve society well, and that we want science to get right—not just the technological devices it spawns but also concepts, measures, data categorisation and curation, and the many other things we mention in the Preface. Moreover, science couldn't get its explanations and theories right if it didn't simultaneously get these other productions right as well. That's why in this book we set out to investigate not 'How does science

---

[15] See https://www.nap.edu/catalog/11876/science-evolution-and-creationism.
[16] See https://sciencecouncil.org/about-science/our-definition-of-science/.
[17] Much of what follows is taken directly from Cartwright (2020).

manage to come up with such good explanations and theories?' but rather 'What goes on in science that secures the reliability of all these different productions? Why are they so likely to do what is expected of them?'.

Theory, at some level or another (see (iii) below for an example of the role of low-level and middle-level theory), always plays a part, but a vast network of other kinds of scientific outputs are required, working as a team, to underwrite the success on any particular scientific endeavour, from designing a satisfaction survey for a new teaching method for your metaphysics module to building the large hadron collider at CERN. So, philosophers and methodologists need to be concerned with how to evaluate *all* the products of science, from theories and models to concepts and measures, studies and experiments, data collection, curation and coding, methods of inference, narratives, devices, technologies, plans and science-informed policies, and more. We do not urge attention to these just because every labourer is worthy of their hire but because these different products of science are mutually supporting. Each successful endeavour in science depends on these being up to the job they are needed for to make the endeavour successful.

The movement for the philosophy of science in practice has long called for broadening attention beyond the propositional knowledge encoded in scientific theories and claims to implicit knowledge and practices. But as we point out in the Preface, we don't want the message to be lost in a focus on practices—where what is underlined is the fact that these are activities, things people do. As we noted, models, measures, data sets, and the like are real products of science. And we expect each to be up to the jobs set it. Paying attention to the rich and diverse display of products that make up scientific activity inevitably enlarges the domain of evaluation beyond what guaranteeing truth to theories can achieve.

Against this background, we urge a refocus: *from the acceptability or truth*[18] *of theories and claims to the reliability of the panoply of outputs the sciences produce.*

1.5 In Defence of Reliability

We should note before proceeding that a concern for reliability does not imply no concern for truth.[19] It may be that what you want from a claim is

---

[18] All we say applies whether we look at truth or at various surrogates for truth on offer (truth-likeness, structural homomorphism, etc.). Throughout, we use 'truth' to cover all these.

[19] You may also wonder how reliability intersects with the notions of reproducibility and replicability that are centre stage now due to the so-called 'replicability crisis' in social science. To say that an

that it be literally true. Or from a model that it accurately represents, say, all the significant causes of some specified phenomenon.[20] You need to assess then, 'How likely is it that the claim or the model can deliver on this demand?'. But an almost exclusive focus on truth distracts effort and attention from the hard job of figuring out what goes into backing up the reliability of all these other kinds of scientific products. Ironically, as we have underlined already and defend under point iii. in this section and further in Chapter 3, without good reasons in support of the reliability of a host of these other scientific products, you will be on shaky ground in assuming the truth of a scientific claim. An easy place to see this is with the need for good concept validation and reliable measures (see the discussion of (iii) and (iv) below for examples). If you don't have these to back up your evidence claims, you don't have firm grounds to support your hypotheses.

We offer five considerations in favour of reliability over truth:

(i)   It is notoriously difficult to nail down what general claim it is that is supposed to be true.

(ii) A great many general scientific principles are not truth apt as they come and when rendered so, they are either false, unwarranted, or of limited utility.

(iii) The reliability of a huge network of other scientific outputs must be presupposed in order to warrant the acceptability/truth of a scientific principle.

(iv) Much of what needs to be evaluated in science isn't a candidate for truth to begin with.

(v)   Reliability invites a crucial question otherwise often overlooked: Reliable for what?

Argument (i). That it is hard to nail down just what claim is being made in scientific contexts is not news. Here we remind you of two familiar sources of the problem: a. 'meaning as use'—the meaning of scientific claims is often (perhaps always) dependent on the network of inferences in which the claim

experimental protocol is reliable for establishing correctly a certain kind of conclusion in settings of kind S is to say that it will *reproducibly* establish a correct conclusion of that kind in S. This does not imply that it will yield the same conclusion if repeated. Even getting the same conclusion on the very same subjects in the very same circumstances is not required. That depends on just what the conclusion is like: see, for example, our discussion of the ideal RCT protocol. It can reliably establish an unbiased estimate of an average treatment effect. But we should expect a different estimate from doing the same experiment on the same subjects in exactly the same circumstances.

[20] On reflection you may realise, though, that all you really want is something far weaker of the kind we discuss: you plan to use the claim or model in various ways with other products of science for a variety of aims and what you want is that those uses be reliable for achieving those aims.

participates, and b. scientific principles are often rendered without quantifiers, as generics, or are supposed to hold only 'ceteris paribus' (*CP*).

Argument (i)a. From Ludwig Wittgenstein (1953) through Wilfrid Sellars (1968) to Robert Brandom (1998), meaning as use has been keenly defended in philosophy. When it comes to science, it is notoriously associated with the views of Thomas Kuhn (1962), Paul Feyerabend (1962, 1975), and Norwood Russell Hanson (1958), who underline the problems it raises for sharing and debating 'the same' claim across different communities of use; and it sparked a host of work on 'trading zones' and scientific 'pidgin' languages when Peter Galison (1997) offered these in remedy to the problems of cross-community debate and consensus.

Think about this in the context of Hasok Chang's (2012) work. For Chang, claims that play an essential role in the successes of past theories are true. For example, according to mid-eighteenth-century phlogiston theory, substances rich in phlogiston, like metals, are combustible. If a metal is 'dephlogisticated', say from burning, phlogiston is released in the form of a flame, and the metal loses its key metallic properties. The phlogiston theory was successful at producing pure air (dephlogisticated air that Joseph Priestley claimed was better for respiration), inflammable air (phlogisticated air), and 'calx' (metal that had lost its shine because of lack of phlogiston), as well as metal (made partly from calx). The theory could also be used to make plain water by combining phlogisticated water and dephlogisticated water.

Claims now discarded (like those of phlogiston theory) that were implicated in a variety of successes are true, Chang argues—true because of their pragmatic usefulness. But that does not mean they are only 'true, pragmatically speaking'. They are true *simpliciter*, Chang maintains, because the pragmatic theory of truth is the only theory that makes sense.

Suppose Chang is right. The fact that these claims are true does not mean they can be lifted from the set of source practices in which they were embedded and inserted into current science. They may be true, but if current science does not provide the same resources to make use of them, *they won't be the same claim*. Conversely, the claims of a more modern oxidation theory, even if they are true, won't be the same claim if set into Priestly's context as they are in the scientific context of the more modern theory. They can't amount to the same things, genuinely employing the same concepts and having the same implications. This undercuts most of the point of trying to establish the truth of claims removed from the tangle of other work that supports them, gives them meaning, and provides the means to make use of them. Particularly important to note is that, when it comes to confirmation, you should be wary of trusting them when elements in that body of work are flawed, for instance unvalidated concepts are used or insufficiently defended measures.

Argument (i)b. A great many of highly useful scientific principles do not come in proper propositional form. So, supposing that only propositions can be true or false, they are not, as they come, candidates for truth or falsehood—they are *not truth apt*. *Equations* are a clear example. We may like to think of them as 'true' but attempts to render them as propositions generally and dramatically diminish their usefulness.[21]

Generics also raise problems. What claim is made in sentences of the form 'X's φ', which are common across the sciences, like 'Neurons transmit signals electrically', 'Democracies do not go to war with democracies', 'People respond to incentives', or 'Limestone reveals its flaws on its surface'? We are inclined to judge some of these as 'true', others 'false'. But so far there does not seem to be a semantics available for them that is really suitable in scientific contexts.

Laws in the form of 'X, *ceteris paribus* [CP]'—ones with CP *clauses* attached either implicitly or explicitly—are also notorious troublemakers. Here there is a semantics on offer that purports to circumvent their troubles. The standard problem is that CP or some such fudge expression is used when X is expected to fail under certain circumstances though holding in others, but it's not known what these circumstances are. The claim then is ill formed. You don't really know what is being maintained.

Michael Strevens (2012) proposes a solution. He considers as an example, 'CP, Printing money causes inflation', which he supposes to be true. But, he notes, 'Under a number of different circumstances, printing money may not affect inflation ([e.g.] the extra currency is hoarded in mattresses rather than spent...)...Because these circumstances are rather diverse, an attempt to specify the economic regularity with any degree of precision will be a daunting undertaking, requiring presumably many clauses, subclauses, parentheses, and footnotes' (653).[22] But these long-drawn-out lists, he argues, are not the way to do it. Rather, economics says merely, 'Ceteris paribus, printing money causes inflation.' Strevens then offers a semantics for CP claims that does not involve a great many clauses, subclauses, parentheses, and footnotes.

Stevens supposes that for these CP claims there is an underlying mechanism that, operating undisturbed, generates the regularity described in the claim. The CP clause refers to this mechanism and its undisturbed operation. He maintains that you can succeed in referring to this mechanism without

---

[21] See Cartwright (2019) for more on this, plus (ii) below.
[22] We follow Strevens's example here, though it oversimplifies the economics, since it illustrates the point adequately both for him and for us.

knowing its structure (in his terminology, the structure is 'opaque' to you). So: supposing Strevens is right, given the existence of the *Printing money mechanism* and the ability to refer to it with the CP clause, 'CP, Printing money causes inflation' is true.

Strevens also notes that economic mechanisms may not be opaque: 'When economists propose a hypothesis such as *Printing money leads to inflation*, they are able to describe to some extent, if not completely, how the mechanism works; such descriptions of course play an important role in picking out the intended targets of inquiry' (672). This mention of 'intended targets' points to our second problem with a focus on truth for general principles.

Argument (ii). Once claims are sufficiently well formulated to be genuine candidates for truth or falsity, they can't do much of the work science puts them to. Consider CP principles as nailed down by Strevens's semantics. These will cover precious little of the 'intended targets of inquiry'. That's because most mechanisms that underwrite CP principles never work undisturbed. Rather they work in real settings, where much else affects outcomes. Printing money, for instance, always happens along with many other fiscal and monetary policies and a host of other economic activities. So money may be printed and yet inflation not ensue. Yet the principle, 'CP, Printing money causes inflation' can be crucial to understanding and modelling what happens in these situations. These complex economic situations are the 'intended targets' for the principle.

Science does not generally use these kinds of CP principles as claims from which to *derive* other claims. Rather, they are used in tandem with much else to *build* models and predictions. Science doesn't need them, then, to be propositions that are candidates for truth. Rather, it needs a network of other scientific products to underwrite their reliability for the purposes to which they are put.[23]

Argument (iii). What other kinds of products should be reliable if a scientific claim is to be judged warranted? What these are will differ depending on the exact form of the principle at stake and on the details of the setting in which it is to be warranted. The reliability of these other products is usually implicitly assumed in warranting a principle. But often items that are new or might be thought to be missing are examined explicitly, as in our example below of a recidivism algorithm. Here we describe just a few of the background products that can be needed, but enough we hope to make it clear that there's no warrant without a great many other reliable scientific products

---

[23] Cartwright (2019) defends a similar claim for equations in physics. Whether or not one disputes her claim about these prized products of physics, it is important not to fall into thinking that physics constitutes science and thus that we need not concern ourselves with the kinds of principles used elsewhere.

to back it up. The first two are fairly widely rehearsed in the philosophy literature, the third is less so.

Ensuring your concepts are meaningful. We start with the obvious fact that you can't warrant a theory whose concepts do not make sense. Just how science does—or should—go about supplying meaning for its concepts is a vexed issue in philosophy, as you will know. Is explicit definition required? If so, what kinds of concepts are allowed in the definiens? What about implicit definition by specifying other claims that the concepts participate in? If implicit definition is assumed, again what kinds of concepts can or should appear in the other claims? Must some at least be measurable or observable? Is measurability of the concept itself required for it to be meaningful? Is some kind of direct causal connection necessary to dub something in the world by the concept? And so on.

We have no views of our own about these questions, nothing useful to add to the discussion. *Except* to point out that whatever your criteria for meaning are, science will need to undertake appropriate procedures to ensure that these criteria are satisfied. And those procedures should be reliable for doing so.

Turning facts into evidence. Next, let's turn to evidence, which is the natural focus when one thinks about warrant in science. But, as we discuss further in Sections 3.2.1, 3.3.5, and 4.1, there is no fact of the matter about whether fact E is evidence for hypothesis H independent of background (auxiliary) assumptions. This is often remarked in philosophy of science but many of its lessons keep dropping out of sight. It is, though, front and centre in much of the literature that argues that successes in science are primarily due to the critical, simultaneously competitive and cooperative nature of science as a social enterprise,[24] and it is also made much of in Michael Strevens's (2020) related arguments that there is no scientific method.

This is notorious in archaeology, where, as Alison Wylie (1989a) reports, '[A]rchaeologists grapple with the problem that the data "do not speak for themselves"' (2). The bits dug up from the ground do not tell you what they are evidence for. In another piece she talks about the ever-revisable 'linking principles necessary for reconstructive inference' (Wylie 1989b, 108). In archaeology, 'middle-range theory' (MRT)[25] is used (inter alia) to label the claims and assumptions that link interpretive claims with claims about the material record that are supposed to support them.

---

[24] For example Longino (1990).
[25] We think these are probably better labelled 'middle-level' since they are middle with respect *both* to degree of abstractness and with respect to range of application.

For some jobs, this MRT can be quite low. For instance, Hartmut Tschauner (1996) discusses Ian Hodder's effort to eliminate theorising about the intentions and understanding of earlier cultures from the basic observations:

> 'When I call some remains on a site a house or a dwelling I must mean that "they" used it and recognized it in a house-like way', Hodder... argues. However, all that is implied in the use of the term "house" is that traces in the material record suggest prehistoric people used the structure in a way we would call house-like. That they thought of the "house" as a "house" is not a necessary part of the argument.   (16)

But already, Tschauner argues, this is to *interpret* the traces in the material record, and that requires theory, even if not very high theory. Tschauner notes:

> Yet this low-level theory seems so completely unproblematic to him that, in the report on his Haddenham fieldwork..., in best MRT parlance, he speaks of "the activities observed" in the excavation. What he "observed" is of course... the statics of the record and the background knowledge that would allow an unproblematic "observation" of the past activities is MRT.   (16)

In case you are inclined to think these problems are peculiar to 'soft' sciences like archaeology, recall from the Preface Di Bucchianico's (2009) story of two warring camps in high-temperature superconductivity. One took the explanatory mechanism to be phonons; the other, magnetic modes. In 2002, new methods rigorously showed a 'kink' in the dispersion curve of reflected photoelectrons. Both camps agreed that the data were correct. But they had wildly different interpretations of it due to the great number of differing assumptions they were also committed to. Each of the two camps claimed that this evidence supported their theory and was incompatible with the opposition's.

Ensuring your validity tests are themselves valid. This problem comes to the fore especially when claims are to be warranted about what a measure or an instrument can do. Consider the discussion by Jennifer Skeem and Christopher Lowenkamp (2016) about whether the Post Conviction Risk Assessment (PCRA) algorithm is generally an accurate and racially unbiased predictor of recidivism.

The PCRA is described as a 'research-based instrument' for predicting the likelihood of an offender's reoffending, one among a number of risk-assessment tools increasingly used across the USA and in the UK to 'inform decisions about the imprisonment of higher-risk offenders, the (supervised) release of lower-risk offenders, and the prioritisation of treatment services to reduce offenders' risk' (1). But there is a great deal of controversy about them, as Skeem and Lowenkamp report:

> The principal concern is that benefits in crime control will be offset by costs in social justice—i.e., a disparate and adverse effect on racial minorities and the poor. Although race is omitted from these instruments, critics assert that risk factors that are sometimes included (e.g., marital history, employment status) are "proxies" for minority race and poverty....In the view of Former Attorney General Eric Holder [in 2014], risk assessment
>
> "may exacerbate unwarranted and unjust disparities that are already far too common in our criminal justice system and in our society. Criminal sentences must be based on the facts, the law, the actual crimes committed, the circumstances surrounding each individual case, and the defendant's history of criminal conduct. They should not be based on unchangeable factors that a person cannot control, or on the possibility of a future crime that has not taken place."
>
> These concerns are legitimate and important—but untested. In fact, Holder specifically urged that this issue be studied. The main issue is whether the use of risk assessment in sentencing affects racial disparities in imprisonment, given that young black men are six times more likely to be imprisoned than young white men [in the USA]....Risk assessment could *exacerbate* racial disparities, as Holder speculates. But risk assessment could instead have *no effect* on—or even *reduce* disparities—as others have predicted....   (3–4)

As a result of their study, the authors arrive at the following general conclusion about the issue, at least with respect to PCRA:

> In summary, PCRA scores are useful for assessing risk of future crime, whether an offender is Black or White.   (32)

To back this up they do not so much call on facts that would readily be seen as evidence. Instead, they produce a number of distinct *tests* that PCRA ought to pass—and, they claim, does pass—in order for this conclusion to be warranted. They begin with a claim about previous validation tests: '[T]he PCRA...was constructed and validated on large, independent samples of federal offenders' (16). They then report on ones from their own research. Here are a couple for illustration:

- An 'inter-rater' validity test for the procedures for assigning the input information for the algorithm: '[O]fficers must complete a training and certification process to administer the PCRA. The certification process has been shown to yield high rates of inter-rater agreement in scoring…'. (17)

- A test looking for so-called 'test' or 'predictive' bias: '[There is] little evidence of test bias for the PCRA—the instrument strongly predicts arrest for both Black and White offenders and a given score has essentially the same meaning—i.e., same probability of recidivism—across groups'. (2)
- Tests using 'a series of bivariate logistic regression models' to determine 'whether an average PCRA score of X corresponds to an average arrest rate of Y, regardless of an offender's race'. (20)

But are the tests that Skeem and Lowenkamp discuss reliable at finding out what they are supposed to? Our point is that the warrant for the general conclusion the authors draw presupposes that these various tests do indeed measure what they are supposed to measure. Here we are not endorsing that the work presented in that paper and elsewhere supports that PCRA can be reliable for predicting recidivism in some (generally not well-enough specified) populations of offenders and is not itself racially biased. Rather, we want to illustrate some of the kinds of scientific products whose reliability is required for conclusions to be warranted, even if they are often not explicitly mentioned or defended as reliable.

As you will see when you get to Section 4.3, Sharon Crasnow's (2020) discussion of the V-Dem measure of democracy (which we discuss again in Chapter 6) is another case where worries about validity are brought to the fore. Measures of complex and controverted social notions ('Ballung' concepts in the language of Chapter 3) like 'democracy' generally are done by producing a set of indices that reflect the variety of different characteristics the concept subsumes. Many of these characteristics are not directly measurable—for instance, degree of freedom of speech or of the press. So they are evaluated, or 'coded', by teams of experts. Which calls for judgements on the part of those experts. Crasnow notes that V-Dem is very attentive to worries about coding accuracy on questions requiring judgement. For instance, they use predominantly in-country experts for coding and they explain, 'Multiple experts (usually 5 or more) code each variable' (16).

So, even if you start out being interested only in truth and warrant, and you are disinclined to think of those in terms of reliability, you can't really have warrant unless lots of other scientific products can be assumed to function reliably.

Argument (iv). Many of the scientific products that need to be assessed are not truth apt. Science creates a huge variety of different kinds of outputs that play different roles in different contexts. Each needs to be able to do the job at hand if you are to rely on its use, and much scientific effort is devoted to ensuring this. These various scientific outputs make a motley assortment. Here are just some, many of which we have already mentioned, in no special order and at no special level of description:

- Theories
- Laws and principles
- Local claims—descriptive and predictive
- Bridging principles
- Models
- Classification schemes
- Methods—innumerably many across the sciences
- Practices
- Concept development and validation
- Measures
- Evaluations
- Devices and materials
- Model organisms
- Statistical analyses and
- Other applications of mathematics, including approximation techniques
- Data curation, classification, and dissemination
- Narratives
- ...

Truth is the wrong dimension along which to evaluate the bulk of these. Whether truth apt or not, for all of these it is crucial to be able to assess: are they likely to do what you want of them?

Argument (v). Reliability immediately invites the question: 'Reliable for what end?' The specification of the end is always important in evaluation, whether it is claims or devices you are evaluating. This is something altogether too easy to overlook in judging principles as true/acceptable. Scientific principles are generally put to a variety of different uses in a variety of different contexts where different bodies of background elements are in place. They will do a good job in some of these but not others, in part because of the issues raised above about meaning as use. So, the stark judgement 'true/acceptable versus not true/unacceptable' will lead you astray much of the time.

The importance of being clear just what purposes are intended has been well rehearsed in the modelling literature. Is the model intended to provide understanding? To provide accurate predictions? Predictions about what?

Should it depict the significant causes of some phenomenon of interest? Perhaps it is supposed to isolate a single cause to study its peculiar effect. Or are you going to probe the model to learn about the world? Just how do you plan to probe it and what do we expect to learn?

It is also brought to the fore in philosophy of science work on measures. Consider the representational theory of measurement (RTM) developed by Duncan Luce and Patrick Suppes (2002) and given a simple articulation recently by Cartwright and psychologist Norman Bradburn (2011). According to RTM, a good measure needs three components plus a defence that the three are appropriate to each other:

- A *characterisation* of the concept to be measured
- A formal *representation* of it (as in a table of indicators or an index)
- Procedures for assigning values to the items measured.

Sometimes the arguments that the components mesh properly are in the form of formal theorems (e.g. *representation* theorems), like the von Neuman–Morgenstern theorem that provides a formal representation of the concept 'utility'. Usually they are informal. The point of these arguments is to show that the representation can do the job of representing the concept characterised, or that the procedures are reliable for ascertaining what values that concept takes in the systems that are measured.

Consider the capabilities account of well-being, developed, in different ways, by Martha Nussbaum and Amartya Sen (1993). Sen characterises well-being as constituted by, informally put, the set of lives worth living available to an individual. Nussbaum specifies ten spheres of human experience that everyone should be above a minimal threshold on. Both stress that the values involved are diverse and cannot generally be ranked in importance or traded against one another. For Sen, there may be no fact of the matter for two individuals as to whose available lives are better; Nussbaum calls improvement in one sphere that is below threshold to advance another sphere, a 'tragic trade-off'.

Both insist that the concept of capabilitarian well-being does not lend itself to providing total orderings across individuals. Yet many attempts to measure it do just that, such as the Alkire and Foster Capability Deprivation Measure and the Krishnakumar 'improved' Human Development Index that Travis Chamberlain (2020) critiques. Chamberlain argues that, despite their care and sophistication, there is no good argument in sight that the procedures specified for these will find out about what they are supposed to despite the fact that they provide total orderings. Those procedures are not warranted as reliable for assessing the capabilitarian well-being of individuals or populations that they are supposed to measure.

Beyond these worries about whether the procedures and representation offered will serve a concept as it is characterised, the measurement literature is equally alert to the issue of whether the concept characterised and its related measure will serve the purpose the concept is intended for. The consumer price index (CPI) as currently measured may serve reasonably well the purpose of estimating the average increase in the price of the designated basket of goods across all the places the goods are available for purchase in the USA, but, as Julian Reiss ([2008] 2016) suggests, not at ensuring that veteran's benefits can secure the same standard of living from year to year, because veterans living on benefits often have little access to large suburban outlets where prices are cheaper and whose prices bring down the CPI.

It is also common to make relative judgements about reliability with respect to how fit for purpose a measure is. Will a poverty measure that counts numbers below an absolute threshold (say $25,750 for a four-person household, as in the USA in 2019) or below a relative one (say, household income below 60 per cent of the average, as in the UK in 2019) reveal the amount of suffering as well as a depth-of-poverty measure that weights individuals/households according to how far below the threshold they are, those farther down getting more weight? And of course, relative reliability judgements are in no way confined to measures; they are made about all of the various outputs that the sciences produce.

Purposes can be specified at a variety of levels of abstraction and generality. This can raise problems for estimating reliability. You may want a measure of inflation that will keep veterans' benefits stable from year to year, or one that accurately estimates the average increase in the price of a designated basket of goods in the USA this year, or you may want a 'good' measure of inflation. We can begin to think seriously about how to defend the reliability—or unreliability—of the CPI for the first two purposes but not readily for the third. The problem is not just that wanting a measure to be reliable as a 'good' measure is exceedingly vague. It also splinters into a variety of different, more concrete purposes, some of which the CPI may be 'good' for and others not. This is also what we claimed about scientific principles at the start of our discussion of reliability: they are reliable in the uses we put them to for some settings for some jobs and not in other settings or for other jobs.

The lesson is that certain kinds of reliability claims lack genuine content because the targeted purposes are underspecified. This creates a danger of inference by false ascent from the concrete to the abstract. This happens when you have good reasons to back the reliability of an item for a particular concrete purpose (as with the CPI for estimating the average increase in the price of the designated basket of goods) but then describe the purpose at a more abstract level (as with the CPI being 'good' for measuring inflation) and act as if the reliability should travel upward from the concrete to the abstract purpose. This can lead to very mistaken expectations since the abstract purpose may cover a host of other more concrete purposes for which the item may not be reliable at all. So it is important to formulate purpose carefully.

Getting clear just what reliability claim you are trying to evaluate is not always easy though. Often it is an iterative process, honing the jobs you expect a scientific product to do as you refine the body of support that it can do the jobs delineated, and vice versa. Sometimes it is only after the fact that you realise that you were focusing on one purpose but implicitly assuming others would be served.

Consider the Vajont dam disaster (which we discuss in more detail in Section 3.4.2), in which an entire town in the Dolomites was wiped away by a gigantic wave of water because a massive limestone landslide fell into the reservoir, the dam resisted the impact, and the overflowing water flooded the entire valley. Engineers had focused on whether a dam built as planned would stand against a range of onslaughts and also on whether the surrounding stone would support it. It seems that it was implicitly supposed, wrongly, that a yes answer implied that human lives in the area would be safe in the face of those onslaughts.

As we discuss in Section 3.4.2, the engineers also relied on a well-supported general principle to tell them about local rock: from the chief engineer, '... the rocks [of the Veneto region] are generally very good [...]. Overall, limestone is honest because it reveals its flaws on its surface',[26] where the generic form of this principle matters. This is not offered as a true universal. In this case, in-depth geological studies were considered unnecessary because the rocks of the area did not raise *visible* concern. Tragically, there was evidence available at the time that this principle was not true of the area rocks. There was clear local knowledge of large limestone landslides up the valley. The engineers were at fault both for neglecting this local knowledge and for not doing a geological study. They were not warranted in taking the generic principle to be locally reliable without more investigation given that lives were at stake.

We've offered five arguments here for the need to refocus from truth to reliability:

---

[26] So the chief engineer Carlo Semenza was reported to have claimed: see Gervasoni (1969).

(i) One understanding of what a claim amounts to in a given setting—its meaning in the concrete—can be 'true', another not so. What's needed then is warrant that one version rather than another can be relied on in that setting to do the job needed there.[27]

(ii) A great many of science's most useful principles are false, unwarranted, or of limited utility if articulated properly as propositions that are truth apt.

(iii) The reliability of a tangle of other scientific products must be presupposed in order to warrant the truth of any principle.

(iv) Much of what needs to be evaluated in science isn't a candidate for truth to begin with.

(v) Reliability focuses attention on the crucial question, 'Reliable for what?' This matters as much for scientific law claims and principles as for anything else.

If you are not yet entirely convinced, Chapter 4 has more to add. There we argue that even if your concern is exclusively with truth and confirmation, confirmation will inevitably lead to the tangle. If you don't have warrant that the essential ingredients in that tangle are reliable, you won't have confirmation.

1.6 Returning to Scientific Method

We have made a long excursion through reliability. But we started out in arms against the idea of scientific method. We are not against scientific method (methods) per se. There's nothing wrong with studying scientific method if you aren't stuck with a focus on theories and claims, truth and acceptability. It is reliability that matters, we argue, even when it comes to claims and theories. And good scientific methods are essential for securing reliability. So the study of scientific method is hugely important—so long as that involves studying what should go on in science to make each and every one of its products, with no privileged preconceptions, up to the jobs they are relied on to do.

---

[27] Howlett and Morgan (2010) present a collection of fascinating studies of what happens to facts (or claims stating facts) as they travel from setting to setting, especially of how their meaning shifts, even for 'well-established' facts.

2

Rigour

Preview

This chapter will look at some typical calls for rigour in science and try to identify what is meant by 'rigour' in them, following the strategy that to figure out what rigour is supposed to *be*, it's a good idea to figure out what job it is supposed to *do* (a tactic we also follow for objectivity in Chapter 3). Thus we will discuss some of the central charms of rigour. This will be followed by a discussion of what drawbacks there are to insisting on rigour. We shall focus for illustration on the biomedical and social sciences and on the use of 'rigorous' evidence in policy since these areas are where the issue has been explicitly discussed in recent times.

We use one extended example throughout—the randomised controlled trial (RCT). That's because the details matter. They give concrete sense to what the abstract ideas we talk about amount to in real practice. Though the details are different from example to example, nevertheless the overall lessons are the same: rigour is a good thing, it makes for greater security. But what it secures is generally of very limited use. This is the takeaway message for the overall thrust of this book. Rigour can't do much in securing the reliability of the vast panoply of scientific products we depend on.

Beyond that we hope to provide you with much else, including reflections on what could be meant by rigour to earn its pride of place, especially in our sample cases of establishing causal conclusions; some distinctions involving *formal rigour, substantive rigour,* and *rigour with certifiable inputs*; an introduction to the credibility crisis in economics if you are not already familiar with it; an account of what causal conclusions about a study population can and cannot be secured with the paradigm of rigour for these—the RCT; reflections on how RCTs can function as part of an evidence base about what happens in populations not studied; and an account of a recent methodology for reasonably reliably warranting causal explanations and predictions about places not studied—though decidedly not rigorously.

The Tangle of Science: Reliability Beyond Method, Rigour, and Objectivity. Nancy Cartwright, Jeremy Hardie, Eleonora Montuschi, Matthew Soleiman, and Ann C. Thresher, Oxford University Press. © Nancy Cartwright, Jeremy Hardie, Eleonora Montuschi, Matthew Soleiman, and Ann C. Thresher 2022. DOI: 10.1093/oso/9780198866343.003.0003

2.1 Setting the Agenda

As we noted in the Preface, we put people on the moon and robots on Mars and we happily submit to laser cataract surgery on our eyes. And rightly so—laser cataract surgery goes wrong only once in around 5,000 times. Clearly, much of what science does is reliable. But what secures that reliability? One answer, or standard part of an answer, is that the outputs of science—devices, theories, measures, and the rest—are reliable because they are the result of rigorous processes. Consider, for example, the US National Research Council's (2002) report, *Scientific research in education*:[1]

- … what unites scientific inquiry is the primacy of empirical test of conjectures and formal hypotheses using well-codified observation methods and *rigorous designs*, and subjecting findings to peer review. (51, emphasis added)
- The extent to which the inferences that are made in the course of scientific work are warranted depends on *rigorous reasoning*. (66, emphasis added)

That's from the United States. But the view is in no way peculiar to the USA. Here, for example, is what the Ministry of Education in New Zealand has to say:

Scientific explanations are accepted as reliable only when they have been subjected to *rigorous testing*.[2]   (emphasis added)

This widespread endorsement of the importance of rigour in securing reliability in science is the reason that we choose rigour as one of our three 'usual suspects' to concentrate on. These endorsements are, we claim, a mistake. There is rigour in science. And it has substantial charm. Because rigour relies on procedural rules, it eliminates reason and judgement, which are taken to be unruly, thereby securing objectivity (in one of the central senses discussed in Chapter 3) and making results more certain. But it has a substantial drawback: you can't do much with it. The kinds of results that can be secured rigorously do not support much at all on their own. They are like short, sturdy sticks, good for jobs where a small rigid piece will help. But

---

[1] Note though, from our discussion in Section 2.2, that, unlike, for example, the EBP movement, they must not mean rigour in any very restrictive sense but, it seems, more in the first sense listed in the *Compact Oxford Dictionary*: 'extremely thorough or accurate'.

[2] See https://seniorsecondary.tki.org.nz/Science/What-is-science-about (accessed 10 October 2020).

relying on rigorous support for the products of science is like resting precious bird's eggs on a heap of sticks rather than, as the Jacana bird does, collecting a mass of different materials and weaving them together, with the sticks, into a nest that will support the eggs even though the nest floats on water.

The first half of this chapter will discuss what rigour means and what it can deliver. We shall use evidence-based policy (EBP) as an exemplar, where the call for rigour has been vigorous and explicit. For instance, look at this description of the UK's costly EBP network of What Works centres from a fourteen-page Cabinet Office account of the network, 'What Works: evidence centres for social policy':[3]

> Together, the What Works centres will cover areas with public spending of over £200 billion, helping to ensure that *rigorous*, high quality, independently assessed research shapes decision making at every level.   (3, emphasis added)

If you ask yourself why you might expect a programme or policy which is offered to you to be reliable to produce the effect desired, there are a number of possible answers. But in the contemporary world of EBP, it is held that the best or perhaps the only good reason for assenting to the conclusion is that it has been rigorously arrived at, with the RCT held up as the gold standard for rigour. We shall look at the RCT to see what it can and cannot deliver, and how rigorously.

2.2 What Can You Do with Rigour?

It is not hard to find calls for rigour and injunctions to be rigorous in scientific contexts. This often happens with no good account of what is meant by rigour in general nor of what, if anything, the features extolled have to do with rigour. For instance, in Jones et al. (1996), a *British Medical Journal* piece titled 'Trials to assess equivalence: the importance of rigorous methods', the word 'rigorous' appears in the title but the word (including cognates) appears in the article only once and not very informatively: 'There should be adequate evidence on the rigour of the trial' (39). What it seems the authors care about is getting it right: 'The material covered in this paper should make it clear that such an approach [the one they don't like, which presumably is not rigorous] is likely to lead to wrong conclusions' (38). The authors also

---

[3] See https://assets.publishing.service.gov.uk/government/uploads/system/uploads/attachment_data/file/136227/What_Works_publication.pdf.

link rigour with reliability: 'A proper appreciation of these issues ensures that when equivalence trials are conducted they reach the scientific standards necessary for reliable conclusions to be drawn' (36).

For another example, consider the Schmitz and Hof (2000) paper titled 'Recommendations for straightforward and rigorous methods of counting neurons based on a computer simulation approach' in the *Journal of Chemical Neuroanatomy*. In this case as well it seems that rigour is associated with getting it right, at least with respect to prediction since the recommended methods are said to 'permit satisfactory predictions', though efficiency, usefulness, and informativeness seem to be involved as well. Schmitz and Hof (2000) say:

> We show that estimates of total numbers of neurons obtained using the fractionator are from a statistical and economical standpoint more efficient than corresponding estimates obtained using the density:volume procedure. Furthermore, the use of two simple prediction methods...permits satisfactory predictions...Finally, we show that assessing the reliability of estimates [using a standard alternative method]...is neither useful nor informative. The present results may constitute a new set of recommendations for the rigorous usage of design-based stereology.   (93)

In cases like this it may be that the specific claims and procedures recommended do help to do the job in view better. But it is not enough to simply claim the epistemic high ground by labelling what you like with the honorific 'rigorous'. You need to do some honest toil to earn that high ground. What is the independent sense that things with the label attached must live up to, to warrant that label? In what sense are the recommended claims and procedures rigorous? Is being rigorous in that sense really what helps? And if so, how much work can that rigorously produced result do? We will propose a sense to the idea of rigour that we think accords well with what is generally meant in scientific contexts. But then, we shall argue, much of what is taken to be rigorously produced isn't, and what is cannot do much work except in combination with a vast array of other products not rigorously producible.

What we say is certainly not news. It seems to us almost self-evident and widely acknowledged. For instance, the US National Research Council's (2002) report, *Scientific research in education* explains:

> Advances in scientific knowledge are achieved by the self-regulating norms of the scientific community over time, not, as sometimes believed, by the mechanistic application of a particular scientific method to a static set of questions. The

accumulation of scientific knowledge over time is circuitous and indirect... and progresses as a result of a not-so-invisible hand of professional scepticism and criticism. Rarely does one study produce an unequivocal and durable result; multiple methods, applied over time and tied to evidentiary standards, are essential to establishing a base of scientific knowledge.   (2)

And later:

Making scientific inferences is not accomplished by merely applying an algorithm for using accepted techniques in correct ways.   (4)

This is expressed more clearly by Stanford Professor Kathleen Eisenhardt (a specialist in strategy and organisation in technology-based companies and high-velocity industries):

The perfectly rigorous study does not exist unless perhaps the study of something obviously trivial.[4]

Nevertheless, you see calls for rigour here, there, and everywhere, not least in the EBP movement. More broadly, anything that smacks of so-called 'qualitative methods' in the social sciences comes under fire in portions of the social science community. In the face of this, qualitative social scientists make valiant efforts to show that their methods also deserve this honorific label, as do others whose work does not lend itself to what some influential players regard as rigorous procedures (for just a couple of many instances, see 'Grand challenges and inductive methods: rigor without rigor mortis' by the editors of the *Academy of Management Journal* (Eisenhardt et al. 2016) or 'Methodological rigor in clinical research' (Yang et al. 2012). So we think the matter is worth a detailed discussion.

Even without establishing a clear definition of the concept of rigour, it is plain that there are reasons that we commonly accept as being good reasons for assent which have nothing to do with rigour. For example, one is for you to say that you trust me to have done whatever it takes to come to the right conclusion. You have reason to rely on me because in your experience I have a very high success rate, or because I am a member of a class of people (e.g. scientists, who have a very high success rate). Trusting like this ducks the question of what I have done to come to the conclusion which I am recommending—including whether or not I have arrived at it rigorously.

---

[4] Email communication between Nancy Cartwright and Kathleen M. Eisenhardt (28 September 2020).

Another case is when you do ask me the basis of my commendation, and you do indeed find it convincing, but my story is very simple. I saw it with my own eyes. That is a good answer. But it would be odd to say that my conclusion has been rigorously arrived at. There are plenty of good reasons for believing which do not amount to applying a process at all (unless you are determined to call everything, e.g. seeing, a process), let alone a rigorous process.

Of course, it may be that such reasons for believing are just no good, or at best only acceptable *faute de mieux*, in the absence of the real thing, rigour. And the language of gold standard as applied to RCTs tends to that conclusion. But we set out to show that there are many good reasons for believing which do not emerge from any rigorous process, and that there is no rigorous way of using or integrating evidence, even if rigorously arrived at, to make decisions, to warrant principles to make predictions, to design measures, to build useful technology, and so forth across the panoply of outputs the sciences produce. Hence the role of rigour in assembling and combining the material required to do these things reliably is limited, even if the rigorous process is impressive and often valuable.

If, then, there are many kinds of reasons in support of reliability, why should we differentiate those supported by a rigorous process or result as being of particular significance? Why might we be tempted to privilege such reasons because of that provenance?

First, a distinction. One meaning of 'rigorous' is 'properly carried out'. So 'I carried out a rigorous inspection of the engine after the accident' may mean conscientious, proper, in accordance with the rules, with the correct process. And it may indeed be a good reason for accepting what I say or what I have designed or built that I have carried out the process properly. Or rather, you may make it a condition of accepting it that I have been rigorous in this sense.

But that is not what the advocates of rigour we have been discussing are concerned with. They are not concerned primarily with proper conformity with the process. They are concerned with when and whether it is right to describe the process itself as rigorous. Of course, we all want the process, whatever it may be, to have been properly carried out. A particular RCT or an observational study, a lab experiment or an ethnographic study can all be done more or less according to the standards set for them. But some of these processes get labelled as rigorous by their advocates and others not. What sense of rigour can make sense of the advocates' claims?

It seems that for a result or a product to be described as having been rigorously arrived at it must be the result of a process which has at least the following features:

a. Nontrivial—not just looking to see.
b. Auditable—you can ask for an account of how in a particular case the process was followed and get a comprehensive answer.
c. Rule-based—you can write out what the process is that you have to follow.
d. Precise, as to
   i. The rules—no 'do what seems reasonable at this point'.
   ii. The inputs—no 'add a pinch of sugar'.
   iii. The output—no 'yep, it works'.
e. Integrable—the rules tell you how to combine the inputs and/or outputs.
f. There is a valid argument (i.e. the conclusion follows from the premises) showing that if all input assumptions and data are correct and you followed the rules correctly, then the output should be as it is supposed to be.

We call these the *formal features of rigour*.

A simple example of a process which meets these tests is adding up a column of three-figure numbers. Suppose there are fifty such numbers—or as many as you like for it to be hard to say what the total is without going through the process of adding. That process meets the above criteria. So it is rigorous, we say.

Which gets us to the charms of rigour. The total which the process of adding produces is certain. It makes no sense to say that these numbers don't add up to 26,736. Anyone who denies that is saying you made a mistake, you failed to input all the numbers correctly (or just didn't follow the rules correctly) or do not understand what adding is. This process produces indisputable results so long as the rules are really followed—after all, those are the rules so the result of following them is the right result for that process. It eliminates anxiety by promising certainty. That is what rigour does. That is its charm. If the input information is correct and the rules are followed, the output is certain, 100 per cent, sure thing.

Of course, few processes fulfil these requirements to the max. The rules may not be entirely precise, or maybe the inputs aren't entirely well-specified, or perhaps there is a good argument that the rules produce correct outputs from correct inputs but the argument is not formally valid. From reviewing a good number of uses of the term, it seems to us that 'rigour' is often stretched to cover cases like that too, so long as departures from the ideal are not too egregious. Nothing of what we argue here requires that we think of formal rigour only in its purest form.

Whether we stick to considering formal rigour in the ideal or allow considerable slippage, it is clear that something more needs to be said about the kinds of input assumptions allowed. After all, *any* method can be made formally rigorous with the right assumptions as inputs. Consider this method: From 'If Programme X works to produce its intended outcomes in Birmingham Alabama' infer 'It will do so in Birmingham England'. This looks pretty shaky on its face. But it becomes deductively valid if you add as input information, 'If programme X works to produce its intended outcomes in Birmingham Alabama, it will do so in Birmingham England'.

This kind of concern about inputs has played a big role in economics since the early 2000s, in the 'credibility revolution', which saw the rise of RCTs, especially in development studies. Economics has long been dominated by mathematics, both at the theoretical level (e.g. game theory models) and at the empirical (econometric modelling). On both sides the work is sophisticated and intricate, theorems abound. There is no want of formal rigour. But deduction is not enough. The assumptions from which the deductions are made must be *credible*. Consider comments from Joshua Angrist and Jörn-Steffen Pishke (2010) in 'The credibility revolution in empirical economics: how better research design is taking the con out of econometrics':

> With the growing focus on research design, it's no longer enough to... [label] some variables endogenous [roughly, caused in the system] and others exogenous [caused outside the system], without offering strong institutional or empirical support for these identifying assumptions.   (16)

In this they are following on from the work of Edward Leamer (1983), of Christopher Sims (1980), and of David Hendry (1980), especially citing Leamer's 'Let's take the con out of econometrics', where he urges the use of sensitivity analysis to combat 'the whimsical nature' of key assumptions. This shows that, as philosopher of economics Julian Reiss (2013) reports, 'Econometricians often find one another's...assumptions incredible' (201).

These economists talk about credible inputs—ones with strong support to back them up. Is this the right demand? You could require instead that the inputs be correct. But correctness is both too weak and too strong for a usable notion of rigour. Too weak because you should like a method for which the requisite inputs can be certified. Even if the input assumptions could sometimes be true, the method is useless if you never have a way to ascertain that. Too strong because people typically talk about methods—like RCTs for establishing causal claims or various methods for adding up—as being rigorous. The method is still rigorous even on the occasions when the inputs are wrong (e.g. the number of days the children attended school was mismeasured in this particular RCT). For this same reason, the demand for credibility is too strong as well.

What's special about the methods that these economists extoll is not that the input assumptions are credible on each occasion of use, but rather that the assumptions are of a *type that can often be made credible*. That is why RCTs are supposed to be the 'gold standard' for causal inference. The salient assumption, which we explain in Section 2.3—orthogonality[5]—is supposed to be made credible in a carefully conducted experiment by random assignment and blinding.

So, besides *formal rigour*, we propose to talk about *rigour with certifiable inputs*. That means that the inputs required are of types you can sometimes take for granted—they are established 'well-enough'—or you have ways to make them so. We propose to stay silent here on the vexed question of what counts as well-enough and how to decide that. Since it is rigour that is in view, one might want to insist that the methods by which the inputs are established are themselves rigorous, both formally rigorous and with certifiable inputs. But that is hopeless. It generates an endless regress. At some point there is no option except to plump for what will be assumed. Recognising this surely weakens the force of the label 'rigour'. Still, it can do work within an epistemic community since in practice where the assumptions are in dispute, it will be hard to make the rigorous label stick.

But what we have said so far is still not enough for rigour to be a useful tool. It is not enough to say correctly that the output is bound to be right supposing the process has been properly followed and all necessary inputs are correct, and you are warranted in expecting the output because those inputs are well warranted. That is useless unless you can say what the output is for. Why did you want to know the total of all these numbers, as calculated by the process called 'adding' with the rules specific to that process? Why did you want to know about the effects of a programme or policy in the population enrolled in the study? You value a rigorous process not just because it ticks the six boxes on the list and can in the right circumstances have outputs you can trust, but because those outputs help you. Even if ours is a good list, all it does is formally to distinguish rigour from the many other processes, or methodologies, which you might use to judge what you can rely on.

---

[5] This ignores the necessity of the other assumptions that are noted in Section 2.3, like the grand metaphysical assumptions that the possible causal processes in the population can be represented by a potential outcomes equation and that there is a true probability distribution over the causes and effects. Perhaps these very large assumptions are ignored because they are thought to be shared with other methods that aim to establish causal conclusions. This is plausible for other statistical methods but does not seem obvious for process tracing methods for instance.

Many of these methodologies (e.g. just looking) are not at all elaborate and tick none of the six boxes. Many are often very effective. The promise of rigour is that if you adopt processes which do tick those boxes, you are (pretty much) guaranteed to avoid certain errors which you would be exposed to if you used less rigorous methods. What those errors are will differ according to what you are doing. A rigorous engine check avoids the danger of worn pistons. The rigour of RCTs avoids the dangers of drawing causal conclusions when you should not due to the interference of confounding factors. All this is of no import unless the outputs that are so firmly secured by avoiding these errors are ones whose reliability really matters to you, either directly in themselves or indirectly by helping to underwrite the security of something further that matters to you. For rigour to matter, it must solve a problem that threatens the reliability of a product you are going to put to use.

We call the requirement that a rigorous process have outputs that matter to your uses *the substantive property of rigour*.

Since we are going to illustrate the charms and harms of rigour using RCTs, we first provide a brief account of what RCTs are and what formally can be concluded assuming they are well-conducted. Of course, when people call for rigour, what they actually want in concrete detail beyond the kind of abstract account we have just offered will be different case by case. We focus on RCTs as a typical example whose virtues and vices are shared by almost anything in science that can lay serious claim to rigour. We go into it in detail because we think the details matter to understanding the real force of our claims. We allow that the RCT method is formally rigorous. But, we shall argue, first, that the promise of certifiable inputs is exaggerated. And the reasons go beyond the obvious that it is hard to guarantee that prescribed procedures are followed properly in practice. Rather, there is an in-principle reason. The certifiability of the inputs is threatened for RCTs because of the problem of the 'unknown unknowns', which paradoxically RCTs are supposed to be in a unique position to deal with according to their advocates.

Second, we shall argue that when it comes to substantive rigour, the kinds of errors that RCTs can help avoid are only a tiny portion of those that you need to worry about for most of the purposes you expect RCTs to help serve. This exemplifies our overall thesis that what you can produce rigorously can't do much on its own. Rigorously backed products need to work in tandem with a panoply of other inputs, the bulk of which cannot themselves lay claim to rigour.

2.3 A Primer on RCTs

An RCT is an experiment that uses a correlation—or more accurately, probabilistic dependence—between a treatment T and a later outcome O in a population to draw causal conclusions about T and O in that population. The causal conclusion depends on the sweeping assumption that where there is a genuine correlation between two factors,[6] there must be a causal explanation. Yet more is needed since we all know that correlation is not causation. T may be correlated with a later O and yet not cause it if T is correlated with other factors that cause O. Such factors are called 'confounders'. For the moment we call the net effect of the confounders C. (In the proof the net effect of confounders is represented by β and w.)

As an easy example, consider the mHealth programme we describe in Section 2.6 that aims to improve child nutrition (O) by introducing new phone technologies at local medical clinics (T). A confounder (C) in this case might be the existence of other programmes similarly aimed at improving nutrition or additional government funding for areas with serious nutrition problems.

Now say that T is *orthogonal to* C if the two are probabilistically independent: Prob (C|T) = Prob (T|C). At its simplest what an RCT does is to try to ensure orthogonality between T and C in the population enrolled in the experiment.

The first stage in trying to ensure orthogonality is random assignment. The population is randomly assigned, half to the treatment group, where everyone receives the treatment, and half to the control group, where no one receives it. Random assignment (supposing the assignment is really followed in an experiment) ensures that T is orthogonal to all confounders at the point of assignment (i.e. C is probabilistically independent of T for that experiment). Recall, however, that probability is an 'infinite long run' notion. This means that you shouldn't expect an equal distribution of C factors amongst the T (i.e. treatment) and not-T (i.e. control) groups in any single randomisation of the study population, but rather that if you repeat the experiment, doing a random assignment again and again, over and over, on exactly that same population with exactly the same characteristics, the sequences of relative frequencies of Cs in T and of Cs in -T will converge to the same limit.

---

[6] That is, a real probabilistic dependence not just a correlation in a finite sample. The notion that matters is probabilistic dependence though we sometimes use 'correlation' for that for ease of expression.

Much can happen to upset orthogonality after assignment of course. So the second stage of an RCT is to guard against that. One common means of achieving this is 'blinding'—of the study participants so knowledge of having the treatment or not cannot affect their outcomes, of those administering the treatment so they do not treat people in the two groups differently, even inadvertently, and the same for those who measure the outcomes and for those who do the statistical analyses.

There are many other possible sources of systematic differences between the two groups that need to be policed as well. Here you encounter the ugly problem of the 'unknown unknowns'. Observational studies can try to achieve real balance of C in a single run of the experiment by explicitly stratifying on different values of C and looking for correlations between T and O within each stratum. If a correlation is found then it cannot be due to confounding of T with C since there is no correlation possible between C and T within a stratum that has a fixed value of C. In the Indonesian clinic case this would mean, for example, pulling out all clinics where some other programme is in place or where a certain amount of funding has been provided by the government. The problem is that you can never manually balance the confounders you don't know about. Instead, RCTs are supposedly designed to do that for you.

RCTs do indeed ensure orthogonality of unknown confounders—though not actual balance—at baseline, which might be a big advantage. But even huge efforts at blinding and policing cannot ensure that no unknown confounders occur post-random assignment that could affect the two groups differently. So, despite early orthogonality, the real processes employed cannot ensure orthogonality after the initial setup and there is, unfortunately, no general way to calculate how close they come to it. That needs to be assessed case by case given all that is known about T and O and the actual circumstances of the experiment.

Suppose, though, that orthogonality were achieved. What then? The nice thing about orthogonality is that it can be *proven* (for more, see Deaton and Cartwright 2018) that if it is achieved then the difference between the average of O in the treatment group minus the average of O in the control group—both quantities that we can measure at the end of the experiment—is an *unbiased estimate of the average treatment effect (ATE) in the study population*. What does that mean?

Start with the treatment effect. The individual treatment effect for an individual patient or pupil, village or district is what difference the treatment— just the treatment alone not added in with the effects of anything else—would make to that individual at the specified time, that is, the value of O the individual would experience if they had the treatment versus the value of O if they did not. This is not directly measurable since no one can both have T and not have T at the same time to enable you to observe both outcomes. The amazing thing about an RCT is that if orthogonality is achieved, the RCT allows you to estimate an average of the treatment effect across the population without being able to measure any of the values that go into that average.

Note 'estimate'. You don't find the actual ATE but you do get an 'unbiased estimate' of it. That the estimate is unbiased does not mean that it is close to correct. How close it is will depend in part on how large the population is. For instance, in the case of flipping fair coins, although the probability of heads in a population of 100 flips = ½ = the probability of heads in 10 flips, the number of heads in the population of 100 flips is more likely to be close to ½ than in the population of 10 flips. Unbiased is a different idea than 'close to the right answer'. Unbiased means that if you did an identical experiment on exactly the same population in exactly the same circumstances indefinitely often, randomising anew each time, the expectation of the observed difference in averages would be the true ATE.

The difference between an unbiased estimate and an accurate one (i.e. one close to the right answer) matters to the debate between advocates of RCTs and those who employ observational studies, and it matters to the issue of concern in this chapter: the advantages and disadvantages of rigour. Unlike an RCT, an observational study can achieve *actual balance*, and sometimes the data are available to do so on the actual population of interest. But of course that is only of *known* confounders. If these were all the confounders there were, the difference in the observational study between outcomes with and without the treatment would give the true ATE for that population. But without any information about the unknown confounders, you've no idea how the observed result relates to the correct one.

You might have reason to think that the influence of unknown confounders is small in a given case, but the argument for this will not have the nice deductive character of the one that shows that the RCT estimate is unbiased supposing orthogonality—it will not be *formally rigorous*. For the RCT, what you measure in a single run of the experiment only gives the right ATE by total chance (though in the right circumstances it will be closer with larger populations), and that ATE is for the population enrolled in the study, which may well not be the population of interest. But, supposing orthogonality (as well as a grand metaphysical assumption seldom called into question about what the causal principles at work are—see the proof in Section 2.3.1 on the 'potential outcomes equation' assumption), it is provable deductively that the single run result is an unbiased estimate of the true ATE. So, the method is formally rigorous for that result. And the randomisation and blinding are supposed to make orthogonality credible when implemented properly, which would make it a method with certifiable inputs. We discuss all this in detail in Section 2.4. Here, we conclude with the deductive proof.

2.3.1 The Proof

The proof begins with what is known as a 'potential outcomes equation', which is supposed to describe the study population. A potential outcomes equation assumes that there is a fixed set of causes that can affect the outcome, written on the right-hand side of the equation, and the causes that can affect the outcome are the same for each unit u in the population (though of course not every unit will experience any or all of them). It also assumes that any particular cause of interest—like the treatment, represented here by x—does not produce a contribution to the outcome on its own but its effects, if any, for u will be moderated by other factors. Here the net effect of those for unit u are represented by $\beta(u)$. All the other factors that can affect the outcome but do not interact with x in doing so are lumped together and represented by w. The symbol $c =$ is to designate that the factors on the right-hand side are a full set of causes of the outcome represented on the left and fix its value. Though philosophically controversial, these metaphysical assumptions are taken for granted in experimental practice—their credibility is generally not in question.

For the proof, it is supposed that there is a proper probability measure over all the variables involved for the population under study and also that a potential outcomes equation of the following form describes the relations between the quantities represented by the variables in that population:

$$y(u)c = \beta(u) \times (u) + w(u)$$

where

- $x(u)$ is the focal cause—the treatment
- $w(u)$ is a function of all other causes except...
- $\beta(u)$, which equals both
  - the difference in $y(u)$ when $x = 1$ minus when $x = 0$ (individual treatment effect) and
  - the net effect of all interactive/moderator variables.

From this you see that the average treatment effect, which is the expectation of the individual treatment effects, is $Exp\ (\beta)$, that is, the study population ATE = $Exp\ (\beta)$.

Next suppose orthogonality of the treatment with the net effect of the other causal factors, defined thus:

$$Exp\ (w/x) = Exp\ (w);\ Exp\ (\beta/x) = Exp\ (\beta)$$

Conditioning on x = 1 and x = 0, taking expectations across the equation, and subtracting yields

$$Exp\ (y/x{=}1) - Exp\ (y/x{=}0) = Exp\ (\beta)$$

Thus, you see that the difference in observed mean outcomes is an unbiased estimate of the study population ATE.

Let us now return to thinking about what is meant by rigour and what it achieves, using RCTs as an example.

2.4 Rigour and the RCT

EBP has been all the rage for almost two decades. It is widely mandated at the international, national, and local levels, and there are huge investments in it. For instance, the UK's What Works Network has nine independent What Works Centres, in areas like health, education, ageing, well-being, and homelessness, plus three affiliate members having to do with youth and higher education, and the Wales Centre for Public Policy as an associate. Their website proclaims that 'Together these centres cover policy areas which account for more than £250 billion of public spending'.[7] The point of the centres is to 'help to ensure that robust evidence shapes decision-making at every level', and they are to do so by, among other things, 'assessing the effectiveness of policies and practices against an agreed set of outcomes'.

This is where RCTs come in. These are widely taken to be the 'gold standard' in rigour for assessing effectiveness claims (i.e. claims about how effective a programme or treatment is for producing targeted outcomes).[8] For

---

[7] See https://www.gov.uk/guidance/what-works-network.

[8] You may wonder what methods are not gold standard. The discussion generally focuses on statistical studies—so called 'wannabe RCTs', population-based observational studies (i.e. without random assignment), and case studies. But there are a host of other sources of evidence of causality that are good indicators though they cannot rigorously establish any kind of causal claim: for instance, Sir Austin Bradford-Hill's (1965) famous nine criteria for causation.

instance, the UK National Institute for Health and Care Excellence (NICE) Guidelines Manual[9] maintains that

> A review question relating to an intervention is usually best answered by a randomised controlled trial...   (199)

Exceptions are allowed.

> There are, however, circumstances in which an RCT is not necessary to confirm the effectiveness of a treatment (for example, giving insulin to a person in a diabetic coma compared with not giving insulin) because we are sufficiently certain from non-randomised evidence that an important effect exists.   (199)

But note the circumstances.

This is the case *only if all of the following criteria are fulfilled*:

- An adverse outcome is likely if the person is not treated (evidence from, for example, studies of the natural history of a condition).
- The treatment gives a dramatic benefit that is large enough to be unlikely to be a result of bias (evidence from, for example, historically controlled studies).
- The side effects of the treatment are acceptable (evidence from, for example, case series).
- There is no alternative treatment.
- There is a convincing pathophysiological basis for treatment (199, emphasis added).

Looking elsewhere, we see that the US Department of Education's What Works Clearinghouse has similar views:

> Well-designed and implemented randomized controlled trials are considered the "gold standard" for evaluating an intervention's effectiveness.[10]

The *British Medical Journal*'s article (Sibbald and Roland 1998) explaining why RCTs are important says the same:

> Randomised controlled trials are the most rigorous way of determining whether a cause-effect relation exists between treatment and outcome...

---

[9] See https://www.nice.org.uk/process/pmg6/resources/the-guidelines-manual-pdf-2007970804933.
[10] See https://ies.ed.gov/ncee/pubs/evidence_based/randomized.asp (accessed 10 October 2020).

And they are thought so useful that you can win a Nobel Prize for promoting them in development economics. MIT and Harvard economists Abhijit Banerjee, Esther Duflo, and Michael Kremer won the Nobel Prize in economics in 2019, because, in the words of the prize announcement Press Release (2019), 'Their research is helping us fight poverty'.[11] How?

> The research conducted by this year's Laureates has considerably improved our ability to fight global poverty. In just two decades, their new experiment-based approach has transformed development economics, which is now a flourishing field of research.

So, RCTS are widely deemed gold standard for warranting effectiveness claims. But is the RCT really formally rigorous—does it satisfy our six criteria (a–f)? Does it have certifiable inputs? And is it substantively rigorous—does it guarantee the avoidance of errors that can get in the way of using the results for something you want? Accordingly, we now look at RCTs and their use to raise four questions:

1. Is the RCT process a formally rigorous process?
2. If it is, what does it establish for you which you would not be able to get otherwise?
3. Are the inputs really certifiable?
4. What part do its outputs play in helping you to do what you want to do?

For the case for the importance of RCTs to evidence-based policy to succeed, the answers to these questions need to be:

1. Yes. RCTs provide
   a. A nontrivial process
   b. Which is auditable
   c. And follows rules
   d. Which are precise
   e. And integrable
   f. And there is a valid argument showing that if all the input assumptions and data are correct, then the output must be as it is supposed to be.
2. RCTs provide an unbiased estimate of the ATE in the study population—supposing that the treatment is orthogonal to (the net

[11] See https://www.nobelprize.org/prizes/economic-sciences/2019/press-release/ (accessed 10 October 2020).

effect of) the other factors that affect the outcome. It is hard to achieve an unbiased ATE estimate by other means.

3. Orthogonality is credible in a well-conducted RCT. Random assignment and careful blinding make it likely.
4. There is a way to use estimates of the ATE from study populations to get reliable estimates of what the ATE will be if you use the treatment in a new population. (This makes the usual supposition that the purpose of using RCTs to assess effectiveness in study populations is to provide robust evidence to shape decision-making about future policies and practices.)

As for the first two requirements. Whether RCTs tick the six boxes which we stipulate as defining the notion of rigour, there is not much to be said. No doubt our list could be longer or better in a number of ways. But we hope that it does a good enough job of distinguishing what are called rigorous methods from other ways of getting to conclusions, and that RCTs would qualify under any such list. So there is not much at stake in agreeing that RCTs are formally rigorous. And if orthogonality is secured, they solve the problem of providing an unbiased estimate by avoiding errors which other methods cannot do. The meat of our doubts about the virtues of RCTs lies in the answers to the last two questions. How readily is orthogonality assured? And are the errors that RCTs avoid the central errors that stand in the way of doing what you generally want: producing reliable predictions about the effects of the programme or treatment in new populations?

On the third requirement. We argue that more is needed than random assignment and blinding to make the input assumptions of RCTs credible. And that identifying and providing warrant for these assumptions requires knowledge that comes from outside the RCT. This causal knowledge cannot typically be obtained by rigorous methods, certainly not by RCTs.

This runs counter to familiar promises that for RCTs the assumptions are satisfied by 'design alone', and in particular a design that among all designs is the single one that can handle those 'unknown unknowns'. Consider, for instance, these remarks:

> It is unethical to construct a study that will not, due to its design alone, produce interpretable findings.   (Gallin et al. 2017, 251)

and

> The Gold Standard is based on the assumption that RCT design alone, regardless of other factors, provides the desired quality.   (Coll et al. 2009, 107)

and

> ...[T]he gold standard of a prospective, blinded, randomized controlled trial,
> where, with a sufficiently large sample size, known but unmeasured covariates
> ('known unknowns') and unknown unmeasured covariates ('unknown unknowns')
> are also balanced between the groups.   (Short and Leslie 2014, 899)

One reason that more is needed than random assignment and blinding is that there are many possible sources of confounding that can happen after randomisation that cannot be controlled by blinding. One well-known possible source of confounding is differential drop-out rates between the treatment and control groups. People drop out of experiments for various reasons, and it may well be that these are correlated with (possibly even due to) other causes of the outcome, or with the chances of improving anyway, or with responsiveness to the treatment. This threatens the assumption of orthogonality.

Orthogonality is the assumption we have been stressing since advocates can often be found saying, falsely, that random assignment secures this assumption. But this is not the only assumption required in the proof in Section 2.3.1 that RCTs can deliver an unbiased estimate of the study population ATE. Besides the deep metaphysical assumptions about the existence and form of the causal principles that govern the outcome for all the members of the study population, there is the practical assumption that every member of the treatment group receives the treatment and no one in the control group does.[12] This can easily fail—practitioners know to watch out for contamination from 'crossover' effects.

Consider, for instance, the question of how to evaluate the effectiveness of the child protection programme, Signs of Safety. We choose this programme for illustration because Cartwright has been at work on this question with Eileen Munro, who is one of a team of three leading the implementation of this programme in the Republic of Ireland (ROI) and in the UK. Signs of Safety is now adopted throughout the ROI as well as in eighty or so local authorities in the UK. What is special about Signs of Safety is that it is relationship-oriented and strength-based. It looks to support and develop features available in the family and environs that can ensure the safety of children.[13]

Good RCTs to evaluate the effectiveness of Signs of Safety in any specific locale will be hard to achieve. Crossover effects are one reason. You can

---

[12] The assumption that $x = 1$ in the treatment group and $x = 0$ in the control.

[13] See https://www.signsofsafety.net/what-is-sofs/.

assign individual social workers, or individual districts, randomly to treatment and control. But social workers interact with each other, sharing information and ideas, and those in the control group will be aware of Signs of Safety. So, post-assignment, social workers may adopt aspects of the intervention despite being in the control wing of the experiment.

There is also the familiar problem that controls won't be using nothing. They must at least be provided the services they would have otherwise. But there is crossover between aspects of Signs of Safety and other social work interventions, so if the other services are using those they may be using a bit of Signs of Safety. Since an RTC won't reveal why a programme works (or doesn't), you won't know if the bit that is important is common in both interventions.

Finally, there is the obvious problem that a control group is not even possible in Ireland because Signs of Safety has been introduced nationally, so all child protection social workers are using it.

These are familiar sources of problems for drawing conclusions from RCTs. Good trialists know to watch out for them, and there are methods on offer to help deal with them. But the methods take a great deal of judgement in their application. This threatens the ready certifiability of RCT inputs. And there are no methods that can deal with the unknown causes that may differentially affect treatment and control groups after assignment.

Happily this warning is now being taken more seriously in the EBP community. For instance, the *British Medical Journal Evidence-Based Medicine* reports that the earlier highly influential GRADE[14] guidelines for assessing strength of evidence for causal conclusions have been revised due to the recognition that

> Study design alone appears to be insufficient on its own as a surrogate for risk of bias [i.e. 'bias' in the sense of correlations with factors other than the treatment that affect outcomes].[15]   (Murad et al. 2016, 126)

This is all to the good. But how do you ascertain the risk of bias? It may be true in many RCTs that it is negligible. None of the familiar problems of

---

[14] As the *BMJ Best Practice EBM Toolkit* describes, 'GRADE (Grading of Recommendations, Assessment, Development and Evaluations)…the most widely adopted tool for grading the quality of evidence and for making recommendations with over 100 organisations worldwide officially endorsing GRADE' (https://bestpractice.bmj.com/info/toolkit/learn-ebm/what-is-grade/).

[15] Though note that the discussion there focuses on 'methodological limitations of a study, imprecision, inconsistency and indirectness' with examples of widely acknowledged types, like failures of 'allocation concealment and blinding', not other kinds of on-the-ground differences that we might worry about in real educational trials.

drop-outs, crossovers, and the like afflict this particular study or they have been properly taken account of, and no confounding has occurred from unknown causal factors. But that won't be easy to establish 'well enough'. After all, it is hard to police factors you don't know about. We are happy to acknowledge that in many cases it may be true that these problems are not serious threats and that in some of these cases there is good warrant to back that up. But that warrant is going to require a whole lot more work than just doing the RCT well.

On the fourth requirement for securing the importance of RCTs for evidence-based policy. There are some important points to be made at the outset. It is now widespread in education, development studies, child welfare, and elsewhere to caution, 'Context matters'. The programme or treatment may have worked (and no need to be fastidious about what that means) in other places. But will it work here? That depends.

Sometimes it seems easy to deal with the effects of context, just because you are confident that it doesn't matter very much. Very large numbers of medical interventions take place on the assumption that it will work as well for him here as it did for her there.

It is hard to say why and how and when such optimism is justified, either in general (aspirin usually works) or in particular (aspirin will work for her). But even if you disregard, and were right to disregard, the doubts expressed above about the evidence for the ATE in study populations, optimism about the effectiveness even on average of this intervention in this new population, even if justified by a good argument, will not be justified by the application of a rigorous process that gets anywhere near meeting the six tests (a–f) for formal rigour, nor to identifying and neutralising the risks that caution moving from the study to the new population.

So, there is a serious problem for taking RCTs to be substantively rigorous, the problem of what is misleadingly called 'external validity'. The term 'validity' is commendatory. It suggests that the conclusion (that, e.g. this treatment will work in this new school or with this new patient) might follow in a particular way, *validly*, from the study results, such as the ATE estimate. And in the same way, *validly*, as we established the ATE estimate. Even if you want to say that the ATE conclusion has been arrived at rigorously, and even if you accept that in a particular case the same treatment will work in your school, there is, we say, no hope that the conclusion about your school can or will have been arrived at in the same way as the ATE estimate for the study population (i.e. *validly*), even if we adopt a more generous definition of valid. There may be excellent, almost conclusive reasons to believe that it will work for you, and you may be certain of that. But it would be misleading to describe them as validly established (unless valid just means good), or that the conclusion has been validated, just as there are some excellent ways of coming to very well-supported conclusions that aren't rigorous. But it would be wrong to call any such reasoning to predictions about your school a rigorous process.

We conclude that:

1. RCTs do indeed, according to what we say is a useful definition, meet our formal tests of rigour.
2. If orthogonality could be secured, they provide robust evidence about average treatment effects in the study population.
3. The inputs are not so readily certifiable as widely touted. You cannot, without further reflection as to other causes, accept their central claim that orthogonality is achieved by study design.
4. The central claim even if true does not support the kinds of practical conclusions you probably want to draw through a rigorous argument.

All four of these answers could be qualified by inserting

1. ...rigorous well enough
2. ...robust enough...
3. ...achieved well enough...
4. ...sufficiently rigorous...

So we are not being overly fastidious. But we say that the degree of failure of RCTs to meet these tests is nontrivial. The points we make about the third and fourth claims are not footnotes. We say more than that 'RCTs prove what works' requires footnotes or qualification. It is not pretty much true, or reliable enough. It is pretty much false and unreliable.

That is only to say that RCTs do not provide a rigorously established basis for policy. It is not to say that RCTs tell you nothing. Nor that there cannot be other rigorous or algorithmic or rules-based ways of formulating policy. We don't believe that, but this section has not argued that either way.

2.5 Getting Beyond the Study Population

The problem we raise for RCTs is one shared by all empirical studies. When you study some particular thing or set of things, what you learn is information *about the things you studied*. This information may be helpful in figuring out facts about other things not studied. Indeed, it often is or you wouldn't bother with the study in the first place. But this is not a gift of the study itself. It depends on a host of assumptions about how facts about the things studied relate to facts about other things not studied. Warrant for these assumptions must come from other sources than that study itself.

This is where the problems for rigour arise. The study results may be secured rigorously, but if the same is to be true for the conclusions drawn about things not studied, then all these additional assumptions should be backed rigorously and the method for choosing and combining them must be rigorous as well. This is almost never possible.

There are three straightforward ways to use RCTs to draw similar conclusions about populations not studied—the 'target' populations.

1. Draw the study population from the target population.
2. Generalisation: up–down inference.
3. Transfer: crosswise inference.

The first allows that certain probabilistic conclusions about the target population can be drawn rigorously. But for the second and for the third it is hard to see how serious rigour could be involved even though these are often done very well. (Note though that when they are well done it is almost always, as we argue, with the aid of a virtuous tangle to support the inference.) We'll discuss each briefly.

1.   **Draw the study population from the target population.** If the study population is representative of the target, the study ATE will be the target ATE. You can get a representative sample by drawing the study participants randomly from the target. But this is difficult to achieve, both in practice and in principle. In practice: because this requires some way to individuate or enumerate the members of the target in order to draw some members from it and then some way to pick them randomly. With large targets this is usually done by a sophisticated computer system. It also requires that those drawn in this way are in a position to participate and are willing and able to do so, which is often unlikely. In principle: because things change over time. You draw the random sample from a population and do the experiment at a particular time. So you can learn about what the target is like at that time. But usually you are interested in what the target is like now, after the experiment, and there's no guarantee that what it was like before is what it is like now.

2.   **Generalisation.** Sometimes an RCT, or a host of RCTs, can be used as part of an evidence base from which to infer that the result seen in those RCTs holds

universally, generally or across some specified range. That's the upwards inference. Then you infer downwards that what holds generally will hold in this new
case. It is notorious that the upward inference cannot be done by any method
that is formally rigorous. There are available some formal schemes for inductive
inference and for the testing of general hypotheses. We look at two very simple
versions of these in Section 4.1.[16] These two simple accounts of inference plus all
the more complex versions of them share the same problem when it comes to
rigour: they all require substantive background assumptions that cannot themselves be warranted by formally rigorous methods. In Elliott Sober's words from
Section 1.2, 'The glass is half empty'.

The downward inference can seem easier: from 'All (or generally) Xs are Ys'
to 'This X is Y'. But the ease of the inference is only skin deep. Consider, for
example, a generalisation we discuss in Section 4.1: 'Where "bad practices"
are responsible for malnutrition, changing bad practice to good will bring
about nutritional improvements'. It is going to take a lot of work to give concrete meaning to 'bad practices' and to devising meanings and measures for
'malnutrition' let alone to establish that this new setting is one where bad
practices are responsible for malnutrition. Consider, for example, the differences in meaning connected with various different anthropometric measures
of malnutrition (like age, sex, length, height, weight, and oedema) and with
establishing an acceptable reference group of healthy children for comparison.[17]
Note, too, the troubles our Vajont dam case (that we discuss in Section 3.4)
brings up for relying on anything short of a true universal—even with a well-
established generic that holds widely, you are going out on a limb when you
expect it to hold in any specific setting without good knowledge about the
circumstances of that setting.

3. **Transfer.** The most straightforward way to make the inference that what
holds in a study setting will also hold in a target setting formally rigorous is to
add the assumption that the study and target settings are similar enough in all
the relevant respects for this to be true. But, just as with generalisation, there's
no formally rigorous way to back that up.

The literature in history and philosophy of science is rife with studies
that provide concrete illustration of how little successful generalisation and
transfer owe to rigour. Here instead we will use as illustration our own

---

[16] See also Norton's (2003b) very thorough discussion of these.
[17] For more on some of the problems involved in designing such measures, see Munslow (n.d.).

recommended way to warrant an inference that a programme will be effect-ive in a target setting, which is unlike both generalisation and transfer, using RCTs, if at all, in a very different way.

We start our argument for this method by thinking about the design of technological devices. You can have other reasons in favour of their reli-ability, but a major source of assurance that they will do what you want of them is a good, detailed model of *how* they will do that. By analogy, a good way to improve the reliability of predictions that a programme will have the desired effect in a target setting is to model just how the programme will produce that effect in that setting. Just exactly what is the causal pro-cess by which the causes introduced by the programme are to lead to the desired outcome? And can a process like that really take place in the target setting? After all, if the programme is to produce the desired outcome there must be some causal process by which it does so. You may not choose to model this process and investigate the chances that it can carry through in the target setting, finding it too difficult, or too costly, or whatever. Nevertheless, these are the very matters that will determine whether the programme succeeds. That recommends modelling it as a good way to get your predictions right.

2.6 Tackling the Job of Prediction Directly

What would such a model consist in? In the policy and programme evalu-ation literature, models of these kinds are called 'theories of change' (ToCs). There are many options on offer: for example, the 'Context-Mechanism-Outcome' models recommended by 'realist' evaluators, pioneered by Pawson and Tilley (1997). For concrete illustration, we discuss the one that Cartwright et al. (2020) have pieced together from various offerings by both philosophers and evaluators—a causal-process-tracing theory of change: a process-ToC (pToC) for short. Figure 2.1 provides an example. You see that it registers a number of different features. We'll describe these only briefly since our main point is to use this as an example to illustrate how little rigour can contribute. These features are:

1. The significant steps by which the programme is to produce the outcome.
2. The casual principles by which each step is to lead to the next. These are not actually pictured in the figure but are meant to be listed along-side. They should be a big help in figuring out the next three ingredients.

3. Support factors. The process-tracing theory of change supposes that the causes that are highlighted in typical causal principles are seldom if ever enough on their own to produce the indicated effect. Rather, to do so they need to operate in tandem with other factors—support factors. (Recall that in Section 2.3 these are represented in the potential outcome equation by βs.) In Figure 2.1, all the factors that together are taken to be enough to produce the next effect in the sequence[18] are enclosed in a box. When necessary support factors are missing and cannot be substituted for, the causal process cannot carry through as expected.

4. Derailers. These are anything that can intervene and stop or diminish a process once it is in train. These are represented by jagged ovals.

5. Safeguards. Safeguards can be thought of as 'walls' that prevent derailers from intruding. When safeguards cannot be implemented against every likely derailer, programme designers should be wary of predicting programme success.

The process-tracing theory of change for how the programme evolves provides a framework for categorising the kinds of evidence that, all told, can make effectiveness predictions reliable. This includes evidence that

- The causal principles that are supposed to work at each step can be called into play and at the right time.
- At each step the support factors will take appropriate values to allow that step to produce the next.
- At each step derailers will be absent or guarded against.

There is no rigorous way to establish these separate kinds of evidence, nor to assess what they amount to all told. The kinds of information that are helpful are varied; so, too, are the methods that can provide evidence for them. What is clear is that, if you want to avail yourself of this additional evidence—as you should—you will have to employ a great variety of different methods and a good deal of judgement. This kind of theory of change, well done and well-supported, can yield reliable predictions about programme success. It can do so even though none of the pieces, nor their overall impact, are rigorously supported. Conversely, we have never seen any way to do the same job rigorously.

---

[18] More carefully, to produce a contribution to the effect.

Turn now to the process-tracing theory of change diagrammed in Figure 2.1 for a real-life casual process set in motion in Indonesia in North and East Jakarta, Surabaya, Pontianak, and Sikka in 2013: mHealth, a programme that introduced innovative mobile phone applications to support nutrition outcomes where there were issues with manual growth monitoring. The one we'll focus on is that incorrect categorisation of children's weight as normal was leading to children being missed from being offered nutrition support services. In 2013, World Vision Indonesia, working with MOTECH Suite, designed a mobile phone-based application to address this, among other challenges hindering nutrition service delivery in Indonesia. (The key for Figure 2.1 is too long to reproduce here in its entirety. We include the key for the first couple of steps below. For the entire thing, see the original paper.)

Let's look at the first couple of steps in this process-tracing theory of change to illustrate. Here is the key for them:

> Box 1: mHealth is administered
>
> Box 1′: Community health workers are mandated to use it
>
> Box 1″: District workers are mandated to monitor, curate and respond

Principle 1, 1′–2: Health workers tend do what they can in their clients' best interest.

Support factors:

1a. Community health workers have the capacity to use mHealth.
1b. Community health workers agree that using mHealth is good for their clients.
1c. Mothers and children attend the community health clinic on a regular basis.

Derailers:

1′a. External pressure to not perform the task or other priorities prevail.

> Box 2: Infants' weight data is recorded in phones

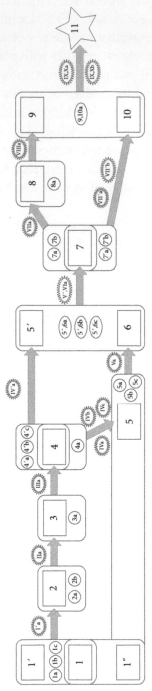

Figure 2.1 A pToC for mHealth, a mobile phone app nutrition programme in Indonesia

Source: Cartwright et al. (2020).

Principle 2-3: mHealth technology does accurate calculations of growth status.

Support factors:

2a. mHealth technology is well designed for the job.
2b. Community health workers input the correct data in the correct format.

Derailers:

IIa. Technology fails to operate.

> Box 3: Accurate classification of infants' growth status by mHealth technology

The two causal principles invoked are typical for this kind of enterprise. The causal principles that programme design and prediction rely on are generally 'middle or low level'—neither high theory meant to obtain everywhere nor so narrow as to hold almost only for the case at hand, and they often come in the form of generics, without quantifiers or explicit range of application. They are what J.S. Mill labels 'tendency principles'. They frequently describe dispositions of individuals or institutions that are widespread but don't appear everywhere, and often where they do obtain they need the right setting or stimuli to be called into play. We discuss these further in Section 6.4.

The two are notably different with respect to the kinds of methods needed for warranting them. The first, that perceiving an action to be in their client's best interest disposes health workers to do that action, is common knowledge. The issue is not establishing it but rather establishing that it will be called into play in this case and then not be overwhelmed by competing dispositions. This places the bulk of work on warranting the support factors and derailers that are identified and evidencing their presence or absence. This is a good reminder that, for making and warranting claims like these effectiveness predictions, not all the methods to be used are taught in science classrooms or methodology courses. 'Call on common knowledge' is an important tool in our methodology tool bag. Yet common knowledge, which can be very reliable indeed, generally has no claim to be rigorously established.

The second principle employed does take warranting. It says that inputting a child's weight and other information into the phone app will produce an accurate judgement about the child's growth status. The methods to warrant this include a great deal of theoretical derivation and use of local knowledge about what the right answer is and about the mechanism by which the app does the calculation, as well as trials under various conditions that it works properly, and more. Not all these methods need be employed in the immediate defence of the expectation that the causal process will carry through from start to finish. Sometimes it can reasonably be taken on faith that they have been employed, and employed properly, by someone else somewhere else. Still, that should not blind us to the recognition that they (or a good substitute for them) are all necessary if the prediction is to be genuinely warranted. Some are of immediate importance though. It would generally be foolish to send phones that have been shipped from abroad out to community health centres without some method that checks that they work properly. The overall point, though, is the same as with the previous principle: little of what we need to establish can be done with methods that can easily claim the honorific 'rigorous'.

Beyond the principles, a host of other things pictured in Figure 2.1 need to be considered—all the support factors, derailers, and safeguards. Why can you assume that the community health workers are able to use the apps? Perhaps because of background knowledge that they are accredited nurses and also use equipment like this all the time and that they have the time to use the phones properly. But in Indonesia that was not the case. The community health workers were for the most part untrained volunteers. Perhaps, then, because there are records that they have participated in a training course, or from questionnaires, or from test results. And so forth.

So, to provide evidence for a claim like 'Adopting this mHealth technology will improve the growth of children attending the village clinics', I propose that you first construct a plausible process-tracing theory of change for the targeted cause/effect pair in the setting under consideration. The theory should contain the five types of information listed above. It is plausible when (a) each step is connected to the previous by a casual principle for which there are reasons in support of the assumption that that principle *could* operate in that setting, and (b) the support factors and derailers at each step are appropriate to that causal principle. Figure 2.1 then shows you what facts should be evidenced as part of a large body of evidence about the overall cause/effect claim. You then want whatever evidence is possible, gathered by whatever methods are appropriate to it, that each and every support factor and requisite safeguard is (or was) likely to be in place and each and every derailer that is not safeguarded against is (or was) likely to be absent. That's a lot of evidence to gather. And little of it can be vouchsafed rigorously. But these are, after all, the facts that have to obtain if the cause is to lead to the effect in the way pictured, and to the extent that any have no evidence in their favour, to that extent our case in favour of the cause/effect claim is weakened. Would that it were easier or could be done more rigorously. But there are few such claims that can be supported in any less demanding or any more rigorous way.

2.7 Final Remarks

Rigour is a virtue. But what can be established rigorously is narrow in scope. We have looked at length at RCTs to show that this is true in one case highly touted as a paradigm of rigour. But there is nothing peculiar about RCTs. There is little substantive rigour to be had anywhere. No method, no matter how formally rigorous, is of use to you if its inputs are not certifiable. But the joint requirement of validity and certifiability will always make the range of use highly limited. The products of science that we rely on are complex and multifaceted. Ensuring their reliability takes more than piling up the kinds of limited outputs you can get rigorously. What validly follows from things you know—things you can take as well-certified—is of limited utility in securing the reliability of these complex, many-faceted products.

Modern science has proved fairly good at securing reliability. But this is not done by rule, not 'rigorously' as in an RCT where, if the procedures are well executed, the requisite premises are likely to hold and the result can be established with a high degree of credibility. Here we echo the US National Research Council's view in *Scientific research in education* (*SRE*), which argues under the heading 'Nature of science':

> Making scientific inferences is not accomplished by merely applying an algorithm for using accepted techniques in correct ways. Rather, it requires the development of a logical chain of reasoning from evidence to theory and back again that is coherent, shareable, and persuasive to the sceptical reader...
>
> (Towne and Shavelson 2002, 4)

Notice the *SRE*'s endorsement of arguing 'from evidence to theory and back', where of course they mean not just once but over and over again. The point is that science does not secure reliability by rigour alone. Far from it. Rigorous results are like the small rigid twigs in a bird's nest. They do some

Figure 2.2 Jacana bird and its nest
Source: Lucy Charlton.

of the job. But a heap of twigs will not stay together in the wind. They must be sculpted, skilfully, with a variety of very different scientific outputs if they are to play a role in supporting the reliability of our scientific products. Just as a bird must weave together a hodgepodge of leaves, grass, moss, mud, saliva, feathers, and so on to build a secure nest, as in Figure 2.2.

3
Objectivity

Preview

What is objectivity and how does it serve science? We set about answering this by addressing the related question, 'What do you expect of science or scientists when you enjoin them to be objective?'. Our answer: they should find out what is to be achieved and what methods will succeed in achieving it. There are many other answers on offer; we are happy to adopt most of them. But we think that more is expected. What is novel about our approach is its emphasis on *finding* and on *aims*.

The emphasis on finding ups the ante.

First, unsurprisingly, there's *finding the methods*. Objectivity is supposed to be conducive to reliability. It is on our account. You are not proceeding objectively unless you are warranted in assuming that the methods you choose can get you where you are supposed to be going.[1] Finding requires more. The right methods to achieve a given aim in a given context are seldom preset. You have to find them.

The demand that you must try to find the right methods doesn't pop out immediately from most of the usual offerings of what makes for objectivity, like following warranted procedures or not being biased. But it is implied by the assumption that objectivity calls for you to be thorough, to be circumspect in your judgements, to avoid jumping to conclusions. This demand requires that you make serious efforts to find out what are the correct procedures to follow in this context or what it is to be unbiased here. You do not just succeed in doing so by chance.[2]

Second is the demand to *find the aims*. This might seem above and beyond the call of the duty to be objective. We urge the contrary. Science matters. What it produces affects human lives and welfare. So aims matter to science.

---

[1] So, objectivity is not always possible. Sometimes it is so important to achieve the goal that you have to try anyway even if you cannot proceed objectively. You just have to bet.
[2] As always, how much effort is required depends on balancing expected costs and expected benefits in the circumstances.

The Tangle of Science: Reliability Beyond Method, Rigour, and Objectivity. Nancy Cartwright, Jeremy Hardie, Eleonora Montuschi, Matthew Soleiman, and Ann C. Thresher, Oxford University Press. © Nancy Cartwright, Jeremy Hardie, Eleonora Montuschi, Matthew Soleiman, and Ann C. Thresher 2022. DOI: 10.1093/oso/9780198866343.003.0004

We illustrate with some clear cases where the connection is immediate—in the design of a dam and scientific advice about badger culling. But the same is true everywhere in science. Part of what earns science the honorific 'objective' is that science tends to get it right when it puts its stamp of approval on a product, whether the product is a dam design or a theory of high-temperature superconductivity.[3] Getting it right means that the scientific product can serve the aims *reasonably expected of it*. Another part of what earns science that honorific is the assumption, just noted, that science is thorough—it doesn't jump to conclusions. So a scientific product has not been produced objectively if there is no warrant available that the product is able to achieve the aims that it should.

What aims are those? What they are may be explicitly stated. But the aims that scientific products are expected to achieve are more often implicitly fixed by context.[4] What goes into context for these purposes? We shall argue that to do the job of delineating the relevant aims (and methods as well) context must include empirical facts, norms, and habits and customs, the products of science and the knowledge that are currently available (both knowledge that and knowledge how), and a shared understanding about expectations.

So, a scientific endeavour is not sufficiently objective, we urge, if it does not make sufficient effort to ensure both that it is working towards the aims that can be reasonably expected in the context and that the methods it employs can be reasonably expected to achieve them there.

We should note at the start that in highlighting finding and aims we only address what objectivity in science can require beyond what is usually on offer. We do not tackle the difficult question of who should be required to meet these demands of objectivity, or when—what scientists, research projects or programmes, disciplines, or institutional bodies in and surrounding science.[5] But as you will see in Chapter 4, we do argue that there is good reason to think these demands will be difficult to satisfy if the right kind of tangle is not available in the background. In that case the possibility of objectivity as we envisage it is under threat alongside the reliability that objectivity is supposed to be in aid of.

---

[3] But, as we urge throughout, this need not mean that science understands correctly what it is doing.
[4] Other philosophers, too, stress the importance of context to objectivity. See, for example, Koskinen (2018), Wright (2018), or Worsdale and Wright (2021).
[5] For instance, they might best be satisfied by a project if some individuals working on it are extremely subjective in their judgements, use their intuitions, jump to conclusions, and also if some are very narrowly focused and blinkered as to all but a very special subset of the aims the work is to serve.

We are trying throughout to offer signposts to make it easy to look independently at the forest and at the trees, especially since many of the trees are interesting in their own right.

To that end, let's start with the trees. This chapter offers a number of ideas that can do philosophic and scientific work independently of their role in our central argument. These include the idea of a loose 'Ballung concept'; two different ways to tighten up such concepts; good reasons that may be in play for not wanting a Ballung concept tightened up; a discussion of 'duty of care' as a model for objectivity in this respect; an account of the Vajont dam disaster that illustrates the interwovenness of values, methods, and facts in science; an outline account of what 'context' could include where context is supposed to help settle appropriate aims; and finally the argument that the usual interpretations of objectivity aren't strong enough for what we want from a notion of objectivity for science. What is needed instead, we urge, is the more demanding notion of *objectivity to be found*, a notion we especially recommend to your attention because it has wide applicability beyond the scientific contexts for which we develop it.

With respect to our general claims for the importance of the tangle beyond the virtues normally cited—represented in our three 'usual suspects'—this chapter aims to show how the common interpretations of objectivity fail to be enough to vouchsafe reliability even in consort with rigour and good method, and that this is equally true of our new notion of *objectivity to be found*, demanding as this notion is.

You may feel that in a sense we paint objectivity as a relatively weak virtue to secure for science. For, we argue, none of the notions of objectivity that we discuss are sufficient to account for the vast amount of reliability that our scientific products exhibit, even when they work in consort with rigour and sound method. In another sense though, objectivity is a strong virtue to have. For objectivity makes demands far beyond reliability. Reliability, as we use the term, is always a two-place relation: product P is reliable to achieve aim A. To develop a reliable product you do not have to take on the difficult job of judging what aims are called for in the context. These are set for you in the product specifications. Reliability is thus judged relative to preset aims. But the kind of objectivity we think science requires—objectivity to be found—does not deal just in preset aims. It is far more demanding, requiring that you find not just the right methods to achieve the aims but that you concomitantly find the right aims.

We turn now to our discussion of objectivity. We build up our understanding of the increased demands we make on scientific objectivity slowly through a series of steps outlined in the next section.

3.1 Our Argument at a Glance

There are a great many different answers to the question 'What is objectivity?'. Section 3.2 of this chapter looks at a variety of these, starting with our own view of three requirements we take to be central to the call for objectivity in making scientific decisions: warranted process, independence, and convergence.

One look at these proposals shows what's wrong with objectivity. It is too complex, too open-ended, too ambiguous. These proposals offer a variety of features that are characteristic of objectivity, often leaving it to context to settle which matters on any given occasion. But invoking context does little to remove ambiguity. What is context? And how does it fix what's supposed to matter about objectivity?

A natural conclusion of these discussions is that the word does no real work, it can lead to confusion, and it can have harmful effects in practice. If so, it should be abandoned. Don't call for objectivity but say precisely what it is you actually want on each occasion.

We recognise the force of these worries and agree that sometimes objectivity can and should be replaced by more precise concepts. But we think the concept does have work to do—important work—and, as it comes, rife with complexity and ambiguity. Thus, we argue, it should neither be universally discarded nor universally retained.

To make this good, we have to be clear about what kind of threat the concept faces and what kind of concept it is that we are dealing with. We do this in Section 3.3. We dub the loose, ambiguous concept of objectivity in use in almost every domain, from everyday conversation to theology, law, politics, and across the sciences, 'objectivity as we know it' (OAWKI). We argue that OAWKI is not a concept to be indiscriminately discarded, as some views suggest. Its loose, ambiguous nature is typical of what Otto Neurath ([1921] 1973) calls *Ballung* concepts. This idea will be explained in more detail in Section 3.3.2. For now, it is enough to note that treating objectivity as a *Ballung* notion allows that it does not admit of one unique definition (even a disjunctive one) that is appropriate in almost all of the contexts where the term is used to do work. There are, we urge, good reasons, both descriptive and normative, not to discard *Ballung* concepts.

We will review some standard reasons, borrowed from non-philosophical literatures, to retain *Ballung* concepts, before turning to our own: preserving the normative force of objectivity as we know it. To explain what we have in mind we liken OAWKI to 'duty of care'. In some contexts it is clear what this expression demands: criteria can be clearly and explicitly laid down. But often, when you have a duty of care, it is part of that very duty to figure out what it requires, case by case. We argue that the same holds for the injunction to be objective—where it is the loose notion of OAWKI that is called into play. This injunction leaves to you the duty to *find* what is objective in the circumstances. So, we argue, OAWKI can be used to enjoin, case by case, what we call 'objectivity to be found' (OTBF).

This, though, raises an earlier worry. We argue that to be objective in the ordinary sense—OAWKI—you need to find what being objective amounts to in the circumstances. What fixes that? *Context.* And just what does context consist of and how does it fix what is objective in the circumstance? Section 3.4.1 provides our attempt at an answer. There we articulate a notion of context that pays attention both to how the purposes of a scientific enterprise are formulated in order for you to be objective at it in the circumstances, and to what practices you are entitled to assume as relevant for accomplishing the purposes. A case study on the Vajont dam disaster in Section 3.4.2 helps to illustrate this view and to draw lessons about why 'finding objectivity' is both an epistemic task and a duty at the same time.

3.2 What Is Objectivity?

There are dozens of proposals on offer about what objectivity consists of, each different. Before surveying a sample of these, we open this section by reflecting on a few reasons why objectivity appears to be such a 'praise word' across so many fields of theory and practice. We will pursue this by using an example that highlights three requirements we take to be central to the call for objectivity in undertaking research tasks.

3.2.1 Objectivity: A Praise Word for Science and Beyond

The demand for objectivity in science will surely require specific practices special to the sciences. Nevertheless, the overall intent should not be very different from all the other places objectivity is required, from theory to law to daily life, otherwise its use as a praise word for science will be misleading. So we begin by asking: What is it about objectivity that makes it so appealing to people who do not attempt to define it and have neither interest in nor knowledge of epistemology and ontology? Why is it that derogatory remarks such as *you are not being objective* carry such power? What is the significance of the comfortable feeling that because your conclusion and the process you and your colleagues have followed are objective you most likely cannot and have not gone wrong? What is so good about objectivity? What do you want it to do for you?

Take the example of a social worker who is to decide the future of a child and what in that case, as a professional, they might see as an objective way of deciding. They are likely to aspire to reaching an agreed decision as a result of independently operating a process that can be trusted.

The first thing you have to do if the process is to be trusted is to establish the facts. Was the babysitter there that evening? Has the father got a history of violence? Often, the facts are hard to find. Nobody can remember what night the babysitter was there. I think anyway the mother is lying about that. The father's history is muddy. And he drinks too much. A lot of unpursued allegations come from the grandfather who doesn't get on with him. You need to be systematic and thorough in your hunt for the facts.

But there is more to reaching a good decision than the establishing of facts. Facts like these only get on the agenda because they are relevant to what you are investigating. They are relevant because they contribute to an account of what happened. They do not just appear before your eyes and without further ado lead to the conclusion. The facts are what the story says you need to find out. Finding the right narrative involves the difficult job of sifting through all the facts to sort those that are relevant. If you don't manage to do that well, what you have done does not warrant trust.

The account that you are interested in constructing is to do with the possibility of abuse. And you have to be clear what facts, if established, support that conclusion. Which requires a clear (enough) idea of what constitutes abuse, if you are to say that as a matter of fact this child has been abused.

We can say in favourable cases that the resulting decision avoided the problems of being unsystematic (proper professional procedures were followed), of prejudice (nobody was trying to please the committee chair), or disagreement (we agreed). And, continuing to give the analysis its due, the accusation *you aren't being objective* works well among professionals as pointing towards not following the kinds of procedures that are required for arriving at a sound decision, having baggage, and hence not agreeing. And it works, when it works, because the group knows without having it all spelled out what kinds of possible errors are in question.

But even if we are to conclude that this sense of objective—*warranted procedures, independence, convergence*—is a useful and effective concept, you must be aware that deciding by being objective in this way is only possible if a large amount of background work is in place, and the bulk of this work will be invisible at the time your conclusions are drawn. Its reliability is not agreed as a result of any objective (in the present sense) procedures. This goes against the well-established (and for long widely accepted) view that if you follow the rules of method you will achieve objective results, and the rules of method cannot be questioned on a charge of not being themselves objectively established.

First, even if you come to be pretty sure of the facts, you have no good general description of what it is to follow a warranted procedure for establishing them. In the case of the father who drinks too much, you may be lucky, and there is evidence which one might call scientific—a doctor's report about his liver, or breathalyser tests. But for the rest, there are only banalities such as 'find and use the evidence which is relevant'. But that so-called procedure is described too generally to tell you what actually to do. So there is no agreement on any such procedure. There is only a (well-founded) assumption that professionals will from their experience be reasonably good at verifying the kind of facts that are relevant to the kind of decisions that they have to make about children.

Second, what is meant by child abuse is (within limits) taken for granted between the professionals. It is assumed that the notion of abuse is part of the agreed process which social workers bring to their work.

But how did this agreement come about? Not as the result of an objective process. The notion of child abuse as we understand it did not always exist. It is not to be found in the Book of True and Relevant Concepts—it was constructed. How does that happen? Child abuse is not just a matter of opinion, but a matter of fact, if not in the sense that Mount Everest is not just a matter of opinion but is a matter of fact. It might be thought to be a problem that the idea of child abuse would not exist if there were not a society to invent it. But that is true of marriage and money, both of which are real enough. They are real because we have robustly defined conditions which have to be met if a relationship is to count as marriage, or if a bit of paper is to be counted as money. A social institution such as marriage does not occur naturally. It cannot be perceived. A society decides, somehow, that marriage consists of a legally sanctioned contract between two individuals, also ratified by religious rule if the individuals so choose, and occurring with prior consent of all parties (at least in some societies), and so on. You know how to identify the facts which have to be verified if this is to constitute a marriage. If you succeed in doing that, the marriage exists, even though it is a construct, even though there is no knockdown reason why it should ever have been created.

So, too, in the case of child abuse, you need to specify robustly the things which have to be present, the conditions which have to be met, if you are to count a concatenation of events and circumstances as child abuse. Try defining child abuse as constituting damage to a child, wilfully applied, over a period … If that is successful you have an agreed characterisation and thus a technique. Then, to conclude that child abuse has taken place, what you have to do is tick the factual boxes. It may be hard. But that is not enough to dispose of the idea of objectivity. You have a clear enough idea of what world you are in if you do tick the boxes—a reliable conclusion through a reliable process carried out by professional people which should generally command assent. So you have what you were after. You have something which you can call objective because you have avoided error by following the right procedures, which includes using the right concepts.

Professionals do not have to have an account of how and why there came to be a shared idea of this concept. But any such account of the how and why would show that it was not arrived at objectively. There is no given set of procedures for arriving at the agreed concepts which are needed to have anything like an agreed procedure for deciding about children.

Consider a (fabricated) account of how another concept came to be agreed. The process of deciding to ban corporal punishment in schools did indeed involve the consideration of easily verifiable facts. The debate only got going because the practice included hitting children, causing them pain, inflicting minor injury, making them cry. These facts were not in dispute and you can see why. What is less clear is the status of the use of terms like 'humiliation' and 'psychological damage'. On the other side, those who favoured the practice argued in terms of deterrence, character formation. Much of the debate was not about the facts at all. It was rather about what should be made of hitting children and causing them pain, whether that constituted cruelty or brutality for example, about what you should or do care about. And differences of opinion about that were not differences between those who were and those who were not in favour of brutality. Everyone was against brutality. What they disagreed about was whether hitting children is brutal.

The meat of the debate is not about collecting more facts—you have more than enough—but about what to make of them, whether they constitute brutality. And 'constitute' does not mean 'add up to', as in the government is constituted of several ministries. It means 'amounts to', what do you make of it. And if we collectively come to the conclusion that corporal punishment in schools is brutal, then we pass very easily to prescription, to saying that it should be banned. You can't be in favour of brutality. You have to say instead 'This is not brutal'. So between the facts and the action lies the hard bit—how to characterise the practice described by the facts. This bit is typically described as being about values, because it is about what matters to us, that children should not be brutally treated. This characterisation requires a tangle of other facts, arguments, beliefs which have to work together. The procedures followed in arriving at this characterisation may well be warranted, but not because they appear in any list labelled 'warranted procedures for settling how to use a concept' and they certainly are not value neutral.

It is impossible to establish a good general description of what it is to follow a warranted procedure for establishing the truth or relevance of a concept, or of arriving at the facts, if by that is meant a general account that covers all areas of deciding, from medicine to economics to playing golf. And that is not what you need. What is needed is narrower. It is a warranted procedure for this particular kind of problem. If problems can be clustered, as, for example, child-welfare problems, then often there are written or unwritten professional standards and rules which exemplify what counts as relevant. So social workers know much of what to look for, because by training, experience, and certification they are members of a profession. And agree about what matters, within limits. And what constitutes child abuse, and how to establish facts in this domain. All that matters is that often enough you can legitimately be criticised for not being objective, as not having subscribed to or having broken some part of the code which in this case we all agree applies to the case before us. So shared experience and training may make us all agree on the fact that the mother is lying, as well as that, if true, that is relevant, and relevant to whether the child has been abused.

So this account of objectivity in terms of warranted procedures, independence, and convergence leaves out a great deal of what lies behind what you have to do when you come to a good conclusion. You can, more or less, now use the notion of child abuse you have arrived at in a way that commands widespread agreement, with justifiable moral and epistemic footing. And this notion may be a very useful one that helps you think in the right way about children's welfare. But when we look under the hood, at the process that gave you those procedures, we find that there is a warren of judgement and values that underwrites the objectivity of the results. There is nothing box-ticking about it. You may end up with a fairly well-understood concept and some widely agreed procedures for finding out about it. But this was not itself the result of a well-defined procedure. You found your way to the notion of abuse by drawing on a tangle of facts, generalisations, theories, common-sense ideas, values which you made work together to give you a notion you could use in practice, as professionals, objectively.

At the beginning of this section we claim that being objective is praised across the board, from theory to law to daily life, and spelled out some of the reasons that makes it so. We think this also helps account for why objectivity is important for science. Similar narratives may be told about categories, definitions, methods, and structures across a broad range of sciences. Take, for example, the field of astronomy which has, as a part of its working methodology, the concept 'planet'. As has been so recently illustrated, however, with the demotion of Pluto, this category is one that rests on the broad agreement of the professionals within the field, an agreement that can be disputed or changed. As with child abuse, there is little disagreement about the properties Pluto possesses—it is a small, icy body in the Kuiper Belt with an orbital period of approximately 250 days and a mass of approximately 1.3×10^{22} kg. Prior to 2006 it was, in many cases, taken to be an example of a planet, but its inclusion in this category was controversial. As with corporal punishment in schools, arguments arose as to how the concept should be applied, not because of any clear factual distinction but because the category itself rests on a network of tangled pieces which define its practical application.

No clear and well-defined procedure existed, nor now exists, by which the category was arrived at. In 2006, in response to mounting evidence that Pluto was just one of many similarly sized objects in the Kuiper Belt and that to include it in the category of 'planet' would require the inclusion of many other objects, the International Astronomical Union set out a new set of rules by which to distinguish the category. Of the three rules, Pluto passed the first two, but failed on the third. It fails to clear its orbit of other objects, sharing its path with a significant number of other (relative to it) massive bodies. Thus, it became a dwarf planet. But again, the conventionality of definitions appears. What counts as 'clearing' an orbit? Even the earth shares its path with other, albeit small, bodies. The answer is predicated not on a clear and objective, well-defined procedure, but on some kind of agreed convergence towards a categorisation that owes more to the astronomers' common-sense intuitions about what types of objects are important enough to us to count as full planets than to an objectively established reference to the necessary properties that planets should exhibit.

The features we have been discussing—*warranted procedures, independence,* and *convergence*—are just three among dozens that are on offer to provide more content to the idea of objectivity. All of these features are like ours: they stand atop large amorphous networks of other work that are essential if these features are to help objectivity get a grip on the world, networks with different components in different relations for different circumstances. Objectivity, it seems, is like rigour and scientific method. It is indeed a feature of scientific activity that can help account for the reliability of the products the sciences create. But objectivity is not a self-contained notion. When we look into it, what we are calling 'the tangle of science' silently underwrites its force.

As noted, we will attempt to make more sense of this notion of a tangle in Chapter 4. Here our job is to make sense of the notion of objectivity. Which of the myriad features on offer comprise it? Some, all, none? Section 3.2.2 includes a brief survey to give you a sense of what these various offerings are like. Our own answer is all of the above, in different combinations, depending on the context. We arrive at that not by trying to figure out what objectivity consists in but instead by considering what jobs it is supposed to do. And just what are those? That too, we argue, depends on context.

3.2.2 Various Attempts to Capture the Meaning

The examples in Section 3.2.1 are an attempt to present what features are at stake when objectivity is your goal, and what problems their clarification bring to the fore. But there are very many other attempts in the philosophical literature at clarifying what objectivity entails. Clarifications are typically achieved by one or the other of two strategies, the *list strategy* and the *precisification strategy*.[6]

3.2.2.1 The List Strategy
The list strategy clarifies what the concept means by listing features that it seems to cover across different domains and circumstances and when applied at different levels and to different referents—processes, people, claims. The result is what we call *kitchen sink objectivity* after the popular Second World War prescription to 'throw everything but the kitchen sink at the enemy'. The kinds of lists on offer fall roughly into two categories.

On one side, objectivity is said to entail, for example:[7]

(1) Grasping the real object (manipulable and convergent Objectivity$_1$), detachment (value-freedom and value neutrality Objectivity$_2$), intersubjectivity (procedural, concordant, interactive Objectivity$_3$).[8]

(2) Faithfulness to facts, absence of normative commitments, freedom from personal bias.[9]

---

[6] On these two strategies, see also Montuschi (2021).
[7] For other lists, see, for example, Lloyd (1996); Fine (1998); Megill (1994). See also Montuschi (2014).
[8] See Douglas (2004) for discussion.   [9] See Reiss and Sprenger (2017) for discussion.

(3) Unique interpretation of a body of data (hermeneutic objectivity), shared method (procedural objectivity), independence of judgement versus publicity of judgement (deliberative objectivity), avoidance of structural epistemic biases (structural objectivity), evidence.[10]

On the other side, the lists include what you should focus on in order to achieve all of the above: rigour, precision, accuracy, replicability of results or inferential robustness, internal validity, reliability, empirical adequacy, measurability, consistency with established body of knowledge, explanatory power and/or explanatory probity, and more.[11]

In a recent offering to the discussion, Sharon Crasnow (2021) groups together some of these features under the headings of theoretical, empirical, and normative constraints, and suggests that the objectivity of a measure consists in its fitting all these different constraints at once. You will hear more about this in Section 4.3.

It is often supposed that all or most of these features are secured by the scientific method. If you all follow the operations of the scientific method and agree that they have been properly followed, then you conclude that the result is objective. That you all agree upon it also becomes an objective matter. Agreement is a direct consequence of having followed the appropriate procedure. Here, all the supporters of the scientific method, from Bacon to Popper, agree that the most straightforward route to objective knowledge is following the instructions of scientific method (in whatever form this is argued to exist).

When the list strategy is endorsed, the kitchen sink concept of objectivity you end up with is inescapably complex, as Heather Douglas (2004) argues. Objectivity is 'an inherently complex concept, with no one meaning at its core, but instead several different senses packed together' (454). Douglas then proceeds with a mapping of the operationally distinct modes that make up the different senses of the concept. Each mode points at 'where to look for the characteristics of objectivity, while the details within the mode tell us what kinds of characteristics to seek there' (454). Equally, each mode includes multiple senses of the term, none of which is logically reducible to one central meaning. There might be connections among modes but '(n)o one concept emerges as core, [...] and no one mode or sense can serve as the surrogate for the others' (455).

---

[10] See Burch and Furman (2019) for discussion.
[11] Examples of these lists can be found in Kuhn (1977); Dupré (1995); Longino (1990); Wylie (2003).

There is always, of course, the problem of how to read the lists. Is the intention that everything on the list must be satisfied, the bulk of the items, or at least some? Though it is up to each proponent of a list strategy to answer this, it looks as if what is commonly supposed in practice is that there should be no egregious failures on any item. Consider (2), for instance. You really can't plead that you were perfectly objective since you stuck to the facts and your work was value free if you admit that it was heavily influenced by personal bias.

3.2.2.2 The Precisification Strategy

The precisification strategy results in *Objectivity Precisifed*. Rather than indulging in complexity, the precisification strategy simplifies the concept by narrowing it down to one concise definition. This can be done in two ways: either by one definition that is to be used universally wheresoever the concept is employed, or alternatively allowing the choice to vary to suit the problem setting but for each problem setting insist on the use of one single concept—no wiggle room to choose what suits, no it-could-be-this, it-could-be-that, but rather get down to the job of saying just what the duty to be objective in this setting requires. Something specific should always be intended.

Old examples of the first are Thomas Nagel (1986)'s 'view from nowhere', or 'objectivity as invariance' as defined by Robert Nozick (1998), or more recently, conducting 'all disputes with reference to the empirical facts alone' (Strevens 2020, xix, 109) and objectivity as 'averting epistemic risks', as advocated by Koskinen (2020).

The second way to approach precisification accommodates a historically sensitive reconstruction of the concept, where objectivity is said to have acquired different meanings depending on when and what it is invoked for. Lorraine Daston and Peter Galison (2007) analyse how the concept has played out in different epochs in the history of science concerned with different epistemic threats. Scientists in different epochs with different epistemic worries meant very different things by objectivity and made very different moral demands of the objective scientist.

In the period of Goethe and Linneus, 'seeking truth is the ur-epistemic virtue' (Daston and Galison (2007, 58). But what is true about the world? Many studies—geology, anatomy, botany, zoology—assumed that behind/within the variable phenomena is a 'true form', much as in Plato's allegory of the cave. This was an Enlightenment ideal of the eighteenth and early nineteenth centuries that required that the scientist be the 'keenest and most experienced observer' (59). It takes the judgement of a diligent genius to sort out the true form behind the deviating instances.

But the nineteenth century feared the dangers of the unruly imagination and the whim of judgement:

> [T]he scientific vices…were a supine acquiescence to intellectual authority, surrender to the equally passive pleasures of the imagination, and insufficient care in the making, storing and sifting of observations.   (225)

The emerging concept was to become 'mechanical objectivity', an ideal aimed at minimising scientists' intervention and preventing knowledge from being tainted by subjective distortions.

Early twentieth-century structural objectivists were 'chiefly concerned with the justification for the claim that science was objective' in the sense that what is objective should be 'common to all thinking beings' (261). So, one could say that being objective meant one thing in the eighteenth, another in the nineteenth, and something yet again in the twentieth.

Of course, even once a very specific characterisation of objectivity is adopted, what satisfying that characterisation consists in can vary from setting to setting. But the underlying imperative that drives objectivity stays the same. Consider Koskinen, for example. She accepts that what risks need to be kept at bay changes with circumstance, discipline, and practice, and so will the mitigation strategy to be adopted.

As an example of what she proposes, she examines one particular risk, collective bias, and considers withholding epistemic judgement as the way to avoid that particular epistemic risk. If in your context that is the risk you must avoid in order to achieve objectivity, then this is a strategy which you may find useful to employ. If I am, as an individual, biased, you can guard against the biases that may arise from that by getting other, presumptively not biased, researchers to do the same experiment or investigation. But if all researchers are biased, as it happened, for example, with the overall ethnocentric outlook of Western anthropology in the nineteenth century, then changing one biased researcher for another will not help. Hence the general recommendation: '[Y]ou will not—even inadvertently—assess the informants' beliefs or the work of early or current scientists by the standards of judgement ingrained into you through culture and education, if you altogether avoid judging these epistemically' (Koskinen 2021).[12] The risks of

---

[12] See https://link.springer.com/article/10.1007/s11229-020-02645-9. Koskinen notes that this strategy does not work universally but only in contexts where other people's ideas are a central part of the object of study.

epistemic bias are avoided by avoiding epistemic judgement—given that, overall, to be objective unexceptionally amounts to 'averting epistemic risks', whatever form these take.

So, the precisification strategy can work two ways. It can pick one meaning out of the many possible and employ it universally. Or it can make different choices across different pronouncements and practices of scientific research, or for the job at hand, case by case, within the domain of the chosen meaning. The second is more flexible since it allows for contextual qualification, something akin to what J.F. Ross (1981) called 'meaning differentiation in use'.[13] You see this exemplified in the history that Daston and Galison tell. In different epochs, objectivity meant different things to different communities of researchers. Each worked with their own 'precisified' version of objectivity.

Either way you do it, precision comes at a price, which C.G. Hempel (1952) refers to as 'alienation'. In commenting on Rudolf Carnap's notion of explication, Hempel claims: 'An explication sentence does not simply exhibit the commonly accepted meaning of the concept under study but rather proposes a specified new and precise meaning for it' (663). Hempel allows that there is a gain achieved by alienation: 'an enormous increase in the scope, simplicity, and experiential confirmation of scientific theories' (701). By making the meaning of a term specific, you gain enough clarity and precision to make claims using that term testable. But—Hempel argues—there is a loss. In so doing you sever it from other contexts and other connotations where the chosen 'precisification' simply does not cut it.

There are certainly favourable cases where notions of objectivity arrived at by either of the two strategies seem just right. If you want to decide, objectively, the length of a table, then measuring may be your method of choice. And we all know how to measure it. It is trivial to ensure that the tape is accurate, that you have not been bribed to give a false reading, that you are using imperial not metric, that you minimise chances of falling into some form or other of epistemic error, and so on. Anyone who then disagrees that the table is six feet long is not offering a possibly interesting variant opinion but is quickly to be left out of the discussion because they just don't understand what is meant by us when we say we are going to measure this table.

But recall our worries about what happens when you look under the hood. Why is measurement a good way to precisify objectivity, or a good choice to

---

[13] Ross explained how in natural languages there are sets of same-term occurrences which carry an internal multiplicity of meanings. 'Meaning differentiation in use' is a general linguistic phenomenon that points at the fact that words, in suitably contrasting contexts, 'differentiate'. According to Ross, differentiation is controlled by analogy rules that overlook the possible adaptations of terms to the circumstances of their occurrence: see Ross (1981).

make for the context at hand from a list on offer? How objective is the route that makes 'measuring' your proxy for objectivity? Ted Porter ([1995] 2020) argues, for example, that quantified procedures, intended to 'convey results in a familiar, standardised form, or to explain how a piece of work was done in a way that can be understood far away' conveniently hide 'a multitude of complex events and transactions' (ix). Their apparent objectivity is the result of strategies of manipulation and communication that should make you question the objectivity of the processes that established quantification as such a highly praised exemplar of objective reasoning.

The route that makes a certain feature eligible for choice, and what makes you choose it in a specific context, is froth with rich, long-tailed tangles of intervening factors, conditions, information, principles, data, assumptions, theories, decisions, and so on, that sometimes—if you are lucky enough to get the right combination—deliver a seemingly objective picture. But there are many situations where you do not have clear, rule-bound, and eventually agreed-upon procedures to help decide what facts and types of knowledge to include from the start, what other scientific products are reliable for what you need them to do, what arguments work in principle, what methods deliver what. Having chosen a precisification or made a selection from a list for the purposes at hand, you can thereafter make defensible claims to objectivity. But you will seldom be able to make the same kind of claim about your choices. Objectivity, like rigour, runs out fairly soon. So it is good that there is so much to supplement it with to back up the reliability of our scientific endeavours.

What the list strategy and the precisification strategy have in common is the idea that there is a way (or different ways) that objectivity can be thought of in the abstract (which does not mean without content), and that it is an idea that can be instantiated in, or adapted to, context—and in enough contexts to justify the abstract case. This prompts us to think more clearly, and more critically, about the role that contexts are asked to play in giving meaning to objectivity.

3.2.3 Reflecting on Context

The relation between objectivity and context has been widely debated in the literature on objectivity—obvious examples being contextual empiricism (Longino 1990), and feminist standpoint theory (see, for instance, Harding 1991, 2004). But these discussions focus on the political and social contexts of knowledge production (particularly to endorse a subjective, perspectival stance on human knowledge) and on the role of contextual values in closing the gap between theory and evidence, rather than addressing how contexts work as mechanisms of meaning formation for the concept of objectivity.

Helen Longino (1990) emphasises that the construction of knowledge is a social process. Arguing for the objectivity of this process inevitably entails engaging with it and describing what regulates it. The presence of background assumptions and values in the evaluation of hypotheses makes objectivity dependent not just on the impersonal rules of scientific method but on publicly acknowledged and interactively shared constraints that secure critical scrutiny, or in Longino's terms 'transformative criticism'. The constraints refer, for example, to the use of recognised avenues (like peer review) for the criticism of evidence, methods, assumptions, and reasoning; the acceptance of shared cognitive values to assess theories and evidence, recognising that these values might change or be criticised in and by the practice of science; and an acknowledgement that intellectual authority should be equally shared among qualified practitioners. The latter constraint is also an appeal to value diversity in the research community, by including (giving equal voice to) under-represented groups (e.g. women).

The idea of context favoured by some feminist epistemology (e.g. standpoint theory) is that of 'location'. The term is used metaphorically to underline that knowledge is always situated (there is no view from nowhere) and comes from socially and politically situated knowers. Social/political location and situated knowledge give rise to the important idea of 'double vision', which Sharon Crasnow (2013), in illustrating Hill Collins's (1986) view, describes this way: a researcher who is marginalised (e.g. because of race, class, sexuality, physical ability, or a combination of either of them) 'is able to see the objects of study both with the eyes of a researcher trained in the discipline and through her own experience from a marginalized social location' (417) (i.e. from her standpoint).

An easy example is that research into heart disease might well have been different if more women researchers had been in place, pointing out that for women the disease progresses differently than in the paradigm case of men. This makes the researcher not neutral, yet, Crasnow argues, not necessarily partial or unobjective. Rather, it allows her to see how many of the specific concepts and methods used in science are problematic. Besides, because of her location, she has access to other and further types of evidence that might prove important to testing or developing hypotheses. This augments the epistemic grounding for objectivity rather than distorting it. But a standpoint is not just a static perspective. It entails achieving awareness of 'the "work" that goes into forging a political, communal (as opposed to an individual) identity' (418). In sum, in its concern to be objective, science should not insist on being apolitical. Science can still be objective if 'located' in appropriately identified contexts of practice—displaying the characteristics variously pointed out by standpoint theory.

We follow these authors and others in supposing that the socio-political dimension of the notion of context is of central importance. But when it comes to the issue of objectivity, our questions go beyond that. Just how do the features of a context and the perspectives you look at it from work, and work together, to give bite to the ordinary loose notion of objectivity—objectivity as we know it? So, we are looking for an account of context that selects out what is needed to succeed at the tasks that objectivity is supposed to serve in the context at hand. To clarify better what is of concern to us we need to say a few preliminary words on what we are up against.

3.3 Saving Objectivity: Preliminary Steps

3.3.1 What Is the Threat?

Why does objectivity—objectivity as we know it—need saving? For the very reasons the list and precisification strategies are used to introduce more refined notions of objectivity: OAWKI is too ambiguous. As you saw in Section 3.2, when you want to grasp what objectivity amounts to, there are a lot of options on offer. Each seems to matter sometime, and none matters all the time to what is really intended. When you want something well defined, it seems that objectivity splinters into dozens of separate pieces.

Objectivity as we know it is a concept that cries out for analysis. Its opponents see it as a lazy option that is of no use. What is it you want when you call for objectivity in science? Okay, perhaps you want different things in different contexts. This is not achieved by a notion that is so loose, so stuffed full of content that it has no real content at all. It might instead be achieved by fixing on objectivity precisified or kitchen sink objectivity. But this takes you dangerously far down the road to no objectivity at all—which is just where Ian Hacking urges us all to go.

Hacking (2015) is outspoken on this. He urges you to dump the notion altogether. No notion of objectivity, he argues, no matter how overarching or inclusive, produces any added value to calling, occasion by occasion, for whatever the specific circumstances require us to deliver on that occasion.[14]

All you need to do is to get down to 'ground level', that is, to work on the details of science and related activities without worrying about 'second-story questions' about what 'to be objective' really means. For instance, Hacking argues, you can, and should, address directly questions such as 'Can we trust medical research when it is funded by pharmaceutical companies?', without concerning yourselves with whether medical research meets the purported standards of scientific objectivity. In general terms, whether a conclusion (or a person, a policy, a method) is good is best addressed without mentioning objectivity. Hence you should not ask what 'it' is, or what 'it' means (21).

Objectivity is not a virtue. It is the proclaimed absence of vice, of many different vices, Hacking claims (26) echoing J.L. Austin's (1962) idea of a *trouser-word*:

> It is usually thought...that...the affirmative use of a term is basic—that to understand x, we need to know what it is to be x...and that knowing this apprises us of what it is not to be an x. But [for objectivity] it is the negative use that wears the trousers. That is, a definite sense attaches to the assertion that something is [objective] only in the light of a specific way in which it might...not have been [objective].   (60, adapted)

As Hacking tells it, 'this is not objective' claims that this is guilty of any one of a number of vices which we cluster together. The word objective is just a convenient shorthand for saying that this has not fallen prey to this vice, or that, and here follows a quite long list, the members of which are similar in some sense. That shorthand does not of course—it is *shorthand*—give any more than a thin account of what has happened. It does not say which vices were avoided, which are in this case particularly significant, nor how they were avoided. It just attaches the certificate [PASSED], like quality control at the end of the production line. And that is useful. You do not need to go behind the shorthand to see which errors might have mattered, or are in question, as you know none eventuated.

What is less useful—and this follows from the notion of a trouser-word—is to say that a conclusion is not objective. That just means [FAILED]. And

---

[14] Taking social work as their case study, Eileen Munro and Jeremy Hardie (2019) also conclude that the term objectivity has become so ambiguous that it should be avoided. Brown (2019) also argues that the concept of objectivity should be replaced by an account of scientific integrity.

then you do need to go into the details. You have to find which vices occurred from quite a long list of heterogeneous possibilities. So, one version of Hacking's suggestion not to talk about objectivity is to say rather that the shorthand *unobjective* is pointless. You have to go behind it and see what exactly was wrong, and you might as well do that directly, by seeing how it is that things went wrong, because, for example, the participants were prejudiced. And it is certainly true that all the hard work in understanding what went wrong can only be done when you start looking at the particular vices. Just saying *unobjective* gets you nowhere.

Let's dump objectivity altogether is not peculiar to philosophers though. For example, in the policy evaluation community, you find calls to replace it by different, more specific notions like confirmability (advocated as part of a wider set of principles) (e.g. Guba and Lincoln 1989; Lincoln and Guba 2007), credibility, dependability, transferability, and authenticity—all of which are seen to increase the 'trustworthiness' of an evaluation. Much of the debate in evaluation was fuelled in the UK, from the 1980s onwards, by anxiety among commissioners of applied research and evaluation who were increasingly relying on qualitative methods for which they were strongly criticised. Discussions focused on what constitutes 'quality' in qualitative research and on how to introduce a language of 'standards' taking a cue from quantitative/statistical criteria such as validity, reliability, and robustness, and on using them to develop qualitative versions or alternatives (Spencer et al. 2003).

There is also a strong call against objectivity coming from the ethical side of the evaluator's discourse. The emergent arguments rely on the ethical obligations of evaluators *not* to be objective, but rather to be explicitly aligned with certain values such as resilience and mitigating climate change and gross inequalities (Seale 1999; Popay 2006; Spencer et al. 2003).[15]

There is one other reason to stop using objectivity that echoes concerns from our discussion of rigour. As with 'rigorous', branding one's favourite method (or claim, model, measure,…) 'objective' is often just a cheap way to claim the epistemic high ground without the effort of honest toil. In the face of this, a reasonable demand is to stop claiming that your undertaking is objective and get down to explaining exactly what is supposed to be so good about it.

Thinking in terms of our two earlier strategies for explicating what is demanded in the call for objectivity, Hacking follows the precisification strategy to its limit, precisifying objectivity entirely away so that the word just disappears from use. Contrary to Hacking, we believe that the word 'objectivity' should not always be abandoned for something more precise in

---

[15] Our thanks to Elliott Stern for making us aware of these discussions in the evaluation literature.

scientific contexts. It has work to do, important work, and so should neither be universally discarded nor universally retained as a petty, if maybe useful, preliminary accusation that something has gone wrong. To make this good, the term needs to retain all the meanings that it is commonly supposed to have in case you need to appeal to any one or any combination of them in any of a wide and undetermined variety of circumstances. This, however, entails reflecting specifically on what kind of concept is fit for this job.

3.3.2 What Kind of Concept Are We Dealing With?

The ordinary loose concept, OAWKI, appears in almost every domain, from everyday conversation to theology, law, politics, and across the sciences, everywhere people engage in enquiry, deliberation, and debate. It is both highly abstract and hugely imprecise. It is what Otto Neurath ([1921] 1973) labelled a *Ballung* concept, as in the German *Ballungsgebiet* for a congested urban area with ill-defined edges. It is an agglomeration, dense, unruly, and context dependent: there is a lot packed into it; there is often no central core without which an item does not merit the label; different clusters of features from the congestion can matter for different uses. Whether a feature counts as in or outside the concept, and how far, is context and use dependent. There seem to be no rules for what facts about a context fix what features are relevant for applying the concept in that context, nor what the mapping is from whatever those facts are to the features that matter in that context.

Objectivity as we know it is indeed loose, but it is not without content. The various accounts of the term we discussed in Section 3.2 were not plucked out of thin air but are grounded in our ordinary sense of the variety of things that might be expected of someone or some process if they are to count as objective.

We shall be using 'duty of care' recurringly as an analogue to illustrate. Thousands of pages of scholarship have addressed the meaning of the term. Consider how it is treated in the California Civil Code. In a summary of the issues, the Peck Law Group reports:[16]

> The general rule in California is that '[e]veryone is responsible...for an injury occasioned to another by his or her want of ordinary care or skill in the management of his or her property or person....' (Civ. Code, § 1714, subd. (a).) In other

---

[16] See   https://www.premierlegal.org/california-law-establishes-the-general-duty-of-each-person-in-his-or-her-activities-to-excercise-a-degree-of-ordinary-care-for-the-safety-oof-others-says-california-personal-injury-lawyer-stevenpeck/#:~:text=Home,California%20Law%20Establishes%20The%20General%20Duty%20of%20Each%20Person%20In,for%20the%20safety%20of%20others.

words, 'each person has a duty to use ordinary care and is liable for injuries caused by his failure to exercise reasonable care in the circumstances....'[17]

Analogously, in English tort law:

> There must be, and is, some general conception of relations giving rise to a duty of care, of which the particular cases found in the books are but instances... The rule that you are to love your neighbour becomes in law you must not injure your neighbour;... You must take reasonable care to avoid acts or omissions which you can reasonably foresee would be likely to injure your neighbour.
>
> (*Donoghue v Stevenson, [1932] AC 562*, 580)[18]

To make use of this, the lawyer needs to clarify 'Who is my neighbour?' Here is a suggested answer (the 'Neighbour test'):

> ...persons who are so closely and directly affected by my act that I ought reasonably to have them in contemplation as being so affected when I am directing my mind to the acts or omissions that are called in question.[19]

So, the lawyer needs to explain what counts as what one ought reasonably to do. Who is this 'reasonable person' who does this?

> The person concerned is sometimes described as "the man in the street," or "the man in the Clapham omnibus,"... Such a man taking a ticket to see a cricket match at Lord's would know quite well that he was not going to be encased in a steel frame which would protect him from the one in a million chance of a cricket ball dropping on his head.
>
> (*Hall v Brooklands Auto-Racing Club [1933] 1 KB 205*, 224)[20]

How does 'reasonable' apply to 'neighbours'—as in employer and employee, doctor and patient, teacher and student, barrister and client, and so on.— and how do we establish that a 'duty of care' applies? UK tort law lists a number of tests:

---

[17] See https://www.premierlegal.org/california-law-establishes-the-general-duty-of-each-person-in-his-or-her-activities-to-excercise-a-degree-of-ordinary-care-for-the-safety-oof-others-says-california-personal-injury-lawyer-steven-peck/.

[18] See https://www.bailii.org/uk/cases/UKHL/1932/100.html.

[19] See https://www.bailii.org/uk/cases/UKHL/1932/100.html.

[20] See https://www.bailii.org/cgi-bin/format.cgi?doc=/uk/cases/UKHL/1932/100.html&query=(title: (+donoghue+)).

- The Anns test: sufficient relationship of proximity or neighbourhood, scope of the duty, the class of person to whom it is owed, the damages to which a breach of it may give rise (*Anns v Merton London Borough Council 1978*).[21]
- The Three Stage test: harm to the claimant reasonably foreseeable harm; sufficient proximity between the parties; fair, just and reasonable, on public policy grounds, duty to impose (*Caparo Industries plc v Dickman 1990*).[22]

Reference to harm and damage entails reference to negligence. The Judicial Council of California Civil Jury Instructions explain:

> Negligence is the failure to use reasonable care to prevent harm to oneself or to others. A person can be negligent by acting or by failing to act. A person is negligent if he or she does something that a reasonably careful person would not do in the same situation or fails to do something that a reasonably careful person would do in the same situation. You must decide how a reasonably careful person would have acted in [name of plaintiff/defendant]'s situation.[23]   (235)

Here is another attempt at clarification, summed up by the Peck Law Group. In the California Supreme Court decision in *Rowland v Christian*:

> [T]his court identified several considerations that, when balanced together, may justify a departure from the fundamental principle embodied in Civil Code section 1714: 'the foreseeability of harm to the plaintiff, the degree of certainty that the plaintiff suffered injury, the closeness of the connection between the defendant's conduct and the injury suffered, the moral blame attached to the defendant's conduct, the policy of preventing future harm, the extent of the burden to the defendant and consequences to the community of imposing a duty to exercise care with resulting liability for breach, and the availability, cost, and prevalence of insurance for the risk involved.'[24]

We choose these examples because they illustrate two standard ways of getting a grip on a *Ballung* notion, explicating what is intended by it without chucking

---

[21] See https://www.bailii.org/uk/cases/UKHL/1977/4.html.

[22] See https://www.bailii.org/uk/cases/UKHL/1990/2.html.

[23] See https://www.courts.ca.gov/partners/documents/Judicial_Council_of_California_Civil_Jury_Instructions.pdf.

[24] Seehttps://www.premierlegal.org/california-law-establishes-the-general-duty-of-each-person-in-his-or-her-activities-to-excercise-a-degree-of-ordinary-care-for-the-safety-oof-others-says-california-personal-injury-lawyer-steven-peck/.

it entirely in favour of a host of other notions that you may choose among to fit the needs of the occasion. The first is by addressing what it means directly (as in the Jury Instructions or in the English tort law) and the second, by giving a sample list (e.g. in the *Rowland v Christian* case, or the Three Stage test). Looking back to Section 3.2.2, you can see that these strategies match the precisification and list strategies employed in the attempts to explicate what objectivity means.

3.3.3 Some Standard Reasons Not to Discard *Ballung* Concepts

Should one go for wholesale precisification of objectivity as we know it, in different ways for different jobs in different contexts, or stick with one loose covering notion? We don't need to start from scratch in thinking about this. There are standard arguments on both sides that can be borrowed from other literatures. We shall look in particular at issues that arise in deciding between common versus context-local measures and indicators, between EU regulations versus EU directives, and between rules versus guidance. These share a number of problems in common and in common with 'objectivity'. We'll discuss these briefly before we turn to our own reason against wholesale precisification—that we want to retain the full normative force of objectivity as we know it.

On the one side sits 'exactly the same for everyone', as in the single loose notion of OAWKI; on the other, 'same idea but different details, built to suit local context'. For example, a common international poverty line like the World Bank's $1.90/day in purchasing power versus the EU social exclusion measure that includes a set of context-local indicators by contrast with the 'guidelines' approach of the Stiglitz–Sen–Fitoussi report (2009)[25] on measures of economic performance and social welfare; or EU *regulations* which, once passed, are directly effective exactly as written in every member state versus EU *directives* that set a goal that all EU countries must achieve but leave it up to individual countries to devise their own laws on how to reach the goal.

Same-for-all measures and legislation ensure comparability and can seem to make for equality and fair treatment across all who are subject to the regulation or rule. But they often do not achieve the desired effect in situ because they are too concrete, as we have noted for precisifications of OAWKI: what

---

[25] See   https://www.economie.gouv.fr/files/finances/presse/dossiers_de_presse/090914mesure_perf_eco_progres_social/synthese_ang.pdf.

is a good indicator of social exclusion in Denmark may not be so in Italy, or the reverse; what counts as satisfying a duty of care in one setting may not achieve that in another; the way to get the same effect of a desired regulation expressed into law in one country may differ from another as different countries will have entirely different legislative landscapes, judicial bodies, and methods of enforcement.

On the other hand, the use of more abstract, less concrete concepts, guidance, and instructions can make interpretation and application difficult—murky—and often prohibitively expensive (e.g. small firms needing expert legal advice to ensure that what they are doing complies with financial guidelines, or guidelines on negligence and duties of care) and it can be, and be thought to be, open to interpretation, in ways not originally intended, as well as to manipulation—choose the measure that makes your poverty figures look better, or the interpretation of requirements that will cost less to implement.

Similar considerations apply to OAWKI, especially in scientific contexts. It seems a good thing if we can find one sufficiently high standard to hold all the different pronouncements and practices of science, like both qualitative and quantitative methods, or the evidence that underwrites a social policy, and the evidence that underwrites installing a bridge, and the evidence that supports licencing a drug. But we haven't found one that doesn't suffer from the standard problems on the other side. What's on offer that has any promise of being reasonably universal is abstract and loose, making interpretation and application difficult, and leaving too much room for gaming the system. Yet, producing longer and longer lists of specific criteria often doesn't work either. It is commonly noted that where tests are set by rule, teachers teach to the test. Science is not exempt from this. Consider the charges that big pharma sometimes conduct randomised controlled trials to the letter of the law but nevertheless produce misleading results.

We think there is more to be said on the side of the *Ballung* notion of OAWKI beyond these standard arguments, however. For none of them brings to the fore the normative force of *this concept*, which we think can play a vitally important role in many situations. We shall argue for this in Section 3.4. Before that, we lay some further groundwork.

3.3.4 Objectivity As We Know It: World-Guided and Action-Guiding

There is a reason to resist attempts to eliminate the *Ballung* concept of OAWKI. Recall Hempel's idea of the 'alienation' of a concept when you replace the wider, looser, pre-scientific meaning of a term with a technical, precise meaning that can be used more easily and accurately for the purpose of evaluating, say, a scientific theory (Hempel 1952, 663, 701).

Though in specific cases there can be specific gains from discarding OAWKI and replacing it with something far more precise, there is also a danger to doing so. Objectivity, like a number of other *Ballung* concepts we use in everyday life and in science—like 'poverty', 'cost of living', 'social exclusion', 'war', and 'democracy'—is simultaneously descriptive and normative.

As Bernard Williams (1986) put it, these concepts are both 'world-guided' and 'action-guiding'. They describe, but they describe things we care about. Max Weber (1949) argued that the social sciences ought to study the things we care about; so they have far less licence than the natural sciences to substitute different, more manageable concepts. The alienation involved in such substitutions may buy the benefits of precision but it can seriously affect the action-guiding force of our concepts.

Let's consider the action-guiding aspects of two examples of *Ballung* concepts which work well for us. First, our ongoing example of 'duty of care'. Teachers, doctors, and other professionals are said to be bound by such a duty. The phrase has real bite—as we have seen, it is used by courts and by professional disciplinary bodies as the basis for deciding matters of liability and misconduct. It appears in legislation. But what counts as the proper fulfilment of that duty depends on the context (e.g. the person to whom the duty is owed). What you ought to do for a disabled old person will be quite different from what is owed to a mentally disturbed child.

Similarly, it is common in commercial contracts for parties to undertake to use their 'best endeavours' to bring about this or that part of what is agreed is to be done. Again, this is serious. It is not vague and unenforceable verbiage. Companies and individuals have to pay damages for not having used their best endeavours. But again, what failure to do so amounts to varies from case to case.

It might be argued that the work done by 'duty of care' and 'best endeavours' for courts and others is precise because they refer you to a list of considerations which you can then apply unambiguously to your particular problem. Just as 'Obey the Highway Code' does not fail for imprecision. But it is hard to see that there is or could be a list of that kind lying behind the concepts of 'duty of care' or 'best endeavours'. There is sufficient commonality between all driving tasks for the injunction 'look in your mirror before overtaking' to be both precise and general as applying to many particular situations. However, it is hard to think of any instruction that is like that if you are talking about what you should do for an old person or a young child, when oil prospecting, or trading futures. And any list would only be an invitation to do not all these things, but rather whichever of them fits this case. You are also invited to add to the list in the light of what occurs to you as you look at this case. Even supposing you have the right list, you can't be told what weights to attach to this or that injunction if, as will happen, they compete.

More likely, such concepts as 'duty of care' and 'best endeavours' are useful only if those who are taking action are competent by reason of training or experience or, more loosely, mindset to see what in the here and now of this problem are the factors to take into account, and how. Besides, any list is only useful (though it may be very useful) for a skilled operator as a list of suggestions. It cannot act as a set of instructions for the unskilled operator, who will not know which items matter here, or how they combine with each other.

We say that 'duty of care', 'best endeavour', and 'objectivity', along with dozens and dozens of other concepts, are both descriptive and normative in Williams's sense—they are world-guided and action-guiding. They are world-guided: the facts about a particular context are crucial in fixing what these abstract concepts amount to in that context. They are normative in a double sense. Of course the injunction 'Use your best endeavours', 'Act with due care', or 'Be objective' is action-guiding. It instructs you to do something. And the context fixes what that is. But there is more to it than that. Yes, you are directed to do whatever it is that the context fixes as constituting being objective, acting with due care, making your best endeavour. But the injunction also lays on you the duty to *figure out* just what that is in the context.

That will take judgement, it may take skill or training, and perhaps natural aptitude. So, as we have urged, some of us will be better than others at fulfilling these duties. Which needs to be taken note of in employing people where duties like this arise. This makes life difficult. But to avoid the many pitfalls of pre-specifying what must be done, there seems no alternative but to demand that the responsible agent must do their best to figure out what duty requires in each separate context.

This returns us to the drive to avoid alienation. If you want to lay on people an obligation not only to do their duty but also to find, from the long list of things that that could consist in, what it does consist in, or even harder, what it consists in in this case that is not even in that list, then you had better avoid alienation and stick with the general, loose, untidy *Ballung* notions: 'duty of care', 'best endeavour', and 'objectivity as we know it'.

The duty to be objective, couched in terms of OAWKI, carries a burden of responsibility, including an epistemic duty. Hard as it is, if you have a duty to be objective here, you must find out—better, figure out—what it is that matters here in these very specific circumstances. What is required in the circumstances is not an objectivity whose character is preset but rather an *objectivity that needs to be found for those circumstances*. That brings us to the concept of OTBF that we expand on in what follows.

Before that, we need to turn to two features commonly associated with objectivity that have come up throughout this chapter but that we have not explicitly addressed yet: value neutrality and strictures against subjective judgement. Thinking in terms of the heavy demands that objectivity makes shows up new complexities in the use of these two requirements.

3.3.5 Objectivity, Subjectivity, and Value Neutrality

Value neutrality and the elimination of subjectivity have both been regularly lauded as essential in science, and both figure commonly among the features that constitute objectivity. Despite this venerable history, since the early 2000s, both value neutrality and the prohibition of judgement have been under attack in scientific discussions, in the philosophy of science, in science studies, and from standpoint theory, including in works by authors of this book (Montuschi 2017a; Cartwright and Montuschi 2014; Cartwright and Hardie 2012).

It is widely argued that science cannot, and often should not, proceed without making judgements, and many of these judgements involve negotiating values. And the values in questions are not only cognitive and epistemic but rather ethical and social (see e.g. Douglas 2000). Science is rife with decision points. The application of theory, the construction of a measure, the blueprint for a new piece of technology, the design and execution of an experiment—none of these can be done by following prescribed rules. Judgement is required.[26] And many of these decision points are ones where different choices have different knock-on effects for things societies care about, but there is no sufficient scientific reason to go one way rather than another.[27]

Look, for example, at the contentious dispute occurring in the United Kingdom over the spread and increase of bovine tuberculosis: are badgers responsible for this, and to such an extent that culling them is necessary and recommended?[28] Scientists were asked to provide the necessary evidence to establish the causes of infection and to make predictions on the effectiveness

---

[26] For an illustration of how these decision points turn into kinds of 'judgement calls', see Montuschi 2017b.

[27] This is much discussed in the recent literature on how and where values enter science. One standard place familiar to readers of this book from Chapter 1 is in the design of a measure. For some good examples of this, see, for instance, Reiss ([2008] 2016) or Atkinson (1998).

[28] For a more detailed description and analysis of this case study, see Montuschi (2017a).

of culling as a measure to prevent spread. Opposite predictions were issued depending on what was *judged to be relevant evidence*. One team's evidence (David King and others) was based on the outcomes of a ten-year experimental trial (randomised badger culling trial, or RBCT). Another team (John Bourne and others) used a calibration of different sources of evidence:[29] a wide natural science grounding in genetic, epidemiological, ecological, and environmental studies of the problem. And they used economic, social, and practical input, such as the technologies of disease management, or the capacity to set up effective culling regimes, as well as animal welfare information. Settling on the right policy prediction requires making decisions about what to judge relevant to achieve the aim and by what means (eradicating cattle tuberculosis), and for Bourne et al. relevant evidence might go well beyond science (Figure 3.1).

Deciding what evidence is relevant requires judgement. Weighing the facts (more or less known, or more or less certain) that matter to a prediction of effectiveness of a policy decision is part of an extended process, from a phase of preparation (identify an issue; frame it; align the relevant components and interested parties) to a phase of execution (formulate the judgement in terms of a decision, or see it through in its possible or actual effects in practice), conducted within a context that informs the decision being made (Tichy and Bennis 2007; Montuschi 2017b).

Or thinking of the consumer price index discussed in Section 1.5, what should be done about sampling suburban outlets in measuring the CPI?

Figure 3.1 Cull the badgers?
Source: Lucy Charlton.

---

[29] Because of the complex nature of the problem addressed it appeared mandatory to Bourne's team to dig more widely than 'simply the delivery of ultimate findings from a scientifically designed RBCT' (Bourne et al. 2007, 32). King's discussion of evidence can be found in King et al. (2007).

Do you design your measure of inflation to benefit pensioners and veterans, on the one hand, or taxpayers, on the other? Sometimes the characterisation of a concept is given, which seems to reduce the number of decision points in designing a measure for it. But that's not entirely right. Given the intended use of the measure, a measure of the concept characterised may have serious predictable consequences. You are then faced with weighing the importance of measuring the concept thus characterised or designing a measure for a concept altered in meaning that does the job better. These kinds of problems were part of the reason for moving from a fixed-basket index for inflation—one that asks how much it costs to buy exactly the same basket of goods year after year—to a cost-of-living index (COLI) that asks how much it costs to buy the same amount of utility as the original basket offered. In commissioned work, switching the aim of your project is often not possible, which is one of the reasons for worrying about commissioned work. As you carry out the work, you may find flaws with the conception of the work contracted for—both intellectual problems and predictable negative consequences become visible. But still, *that* is the work you are contracted to do.

So we face a conundrum. Objectivity is good in science, and value neutrality and the reining in of subjectivity are central features of objectivity. Yet science cannot, and often should not even where possible, be conducted without judgement and without the intrusion of values.

In Section 3.3.4 we discuss the difficulties of finding and discharging the duty of care and the duty to be objective. Both lay a heavy burden on people, and many of us may not be good at them, or good in some circumstances and not in others. You can do your best and still fail where others would have succeeded. This conundrum makes it even harder.

You are expected to use your judgement in science, and you can be faulted if you don't—'What? You just followed the rules without thinking about it?', or 'You must use your subjective judgement, but you must not be subjective in doing so'. But it is also both a common and a fair charge, 'He was too subjective in his judgement'.

If the recommendation is 'Don't be subjective in your judgements', all three of the features we offered in Section 3.2.1 as central to objectivity constrain what you should do. First, convergence with other independently operating scientists: of course convergence is never anywhere near unanimous, especially when you get down to the details. But if your proceedings are out of line with what others would do after thorough consideration, you'd better have a good reason, and the reason cannot be a hunch. Then you are not being objective in your subjective judgement.

This is not to say that you are never right to act on your own personal assessment if you have good reason to think you are in a good position to get it right. A second aspect of objectivity we highlighted at the start is independence. This means that you are to avoid bandwagon bias and groupthink, inappropriate but possibly unconscious effects of power and status relations, the downplaying of quiet voices, and the like. You also have many duties besides the duty to be objective, including the duty to do what you think is best if you are responsible. Duties often conflict. Think about the duty of care a banker has to her clients in selling a mortgage. This can conflict with the duty she took on as an employee to try to make a profit for the bank. Happily for the banker, that conflict has a set resolution. The law dictates that the duty of care dominates. But usually it is a matter left to judgement, and your duty to be objective dictates you are not to be subjective in making this judgement.

What about warranted procedures? You can't just make it up out of whole cloth how you will proceed. There has to be good reasons for what you do. Yet, generally, there will be no preset list that says, 'This is the way to accomplish this goal in these circumstances'. That was one of the points of our discussion of child abuse at the start. So what are you to do? You must be thorough, systematic. You must gather all the facts that are relevant and choose the methods that are likely to achieve the outcomes aimed for. But aims might need to be juggled, as in designing a measure for inflation. And facts don't come labelled 'I am relevant to X'. What is relevant depends on the context—on the empirical facts, on what else is known, on what standards are expected to be met.

The same holds for methods. It would be great if there were set prescriptions for these: if you want to establish a causal relation, use an RCT. But as we saw in Chapter 1, it doesn't work out that way. Questions always have more fine structure. Seldom is the question really 'What's an unbiased estimate of the ATE for this cause–effect pair in this population at this time?'. More than one method is needed to answer the fine-structured questions. They require specific background knowledge if they are to provide reliable results, their reliable use depends on the reliability of other methods and their products, the exact meaning of the concepts and principles you might call on depends on context, and so forth.

All this requires your subjective judgement—but those judgements must be made objectively. The burden is great, and getting it right will take skill, knowledge, good sense, and finesse. Exercising judgement is a complex process; it does not rely just on 'gut feeling'. It has a procedural nature (though sometimes implicitly so) that entails, as noted, a phase of preparation and a phase of execution (as described in Tichy and Bennis 2007). In both stages, exercising judgement is just one call among many. And for all of these, the context where a judgement call is made is of utter importance in justifying one choice over another (for more discussion of this, see Montuschi 2017b).

Similarly, you are to make value judgements but in a value neutral way, which is of course hard to achieve.[30] Still, you are expected to be non-partisan, you are not to be biased, partial, prejudiced, or self-serving. You must have reasons for your choices among values, and some may pass even if other people would have done differently: 'We made the choice in favour of taxpayers because the already struggling middle classes would have been hit hard but the pensioners would suffer only a little with the reverse decision.' Other reasons will get you laughed out of court: 'We decided in favour of the taxpayers because that's good for my old school cronies (or, my employer, my church,…).'

All told, not much more can be offered as advice independent of context than that you are to make judgements that are balanced and fair, that are based on good accountable reasons, and that have given due consideration to all the relevant issues. Finding how to be objective is no easy matter.

3.4 Objectivity to Be Found

After laying down the grounds for our rescue of the concept of objectivity, it is time to turn to the notion that has emerged as crucial to understanding where rescue lies: the notion of context. We have argued that the call for objectivity in a given setting—OAWKI—lays on you the obligation to discover what in that setting constitutes the objectivity to be found there: objectivity to be found. The *context* of the setting is supposed to fix what that is. If that's the job to be done by context, just what must be meant by context and how does it do that job?

3.4.1 What Is 'Context'? How Does It Relate to Objectivity?

To explain our view of context, we will make use of some ideas developed by Hasok Chang.[31] Chang shares with us the fundamental assumption that

---

[30] As old positions à la Nagel (1961) or à la Weber (1949) bear witness. For a summary description of these positions, see Montuschi (2016), 285–7.

[31] For a wider discussion of Chang's view in the context of objectivity, see Montuschi (2021).

evaluating the products of science only makes sense relative to a choice of purposes to which they are to be put. The first idea we borrow from Chang is the recognition that even simple-sounding purposes are generally complex— buried within them are a number of subsidiary purposes that must be served if they are to be accomplished. The second is a distinction between internal and external purposes.

First, an apparently simple aim such as 'weighing with a balance' entails a wide number of associated purposes that secure its achievement, for example certifying the standard weights, which in turn requires ordering the weights from a reliable supplier, or comparing them to a more trusted set of weights, or checking them against certain natural phenomena (e.g. the weight of a certain volume of water at a certain temperature). Each purpose requires a dedicated activity, none of which is simple, nor simpler than the original aim. And there is no 'lowest level of description', as a reductionist picture would suggest. You just decide, for whatever accountable reason, to stop at some point, for the purpose suggested by some analysis you have settled on (Chang, cited in Soler et al. 2014, 74).

Second, aims are of two types: inherent purposes and external functions. If the activity is 'match-lighting', its inherent purpose is to get a match to light, and all the operations, mental and physical, performed within this activity are geared towards that purpose. External functions are more distant aims the activity might serve. Lighting a match might aim at lighting a candle because the electricity went off, setting a forest on fire as an act of arson, or wanting 'to watch and admire the marvellous process that combustion is' (Chang, cited in Soler et al. 2014, 74).

Often the internal purposes are clear, either explicitly or implicitly. You are instructed to be objective in doing X. Doing X is the internal purpose. Figuring the external purposes may take work, and just what they are in the circumstances may be more open-ended. But it is important to consider both internal and external purposes together because that can have a big effect on what scientific products will be reliable for achieving them.

For a concrete illustration involving complex aims, think about the assignment to develop a 'good' measure for inflation. We suppose that you are supposed to 'be objective' in doing so. From the point of view of the representational theory of measurement discussed in Section 1.5 this involves achieving a number of subsidiary aims: providing a good characterisation of inflation, designing a good formal representation for it, choosing procedures for assigning values on the ground, and defending that the characterisation, representation and procedures mesh properly. So, the general aim is supplied but you have to find the appropriate subsidiaries.

And what these are is not arbitrary. They are fixed by the interaction of the aim, the 'context', and your duty to be objective. In this case the context includes the decades of work by scientists and philosophers developing the details of the theory of measurement. One of the features in objectivity lists is 'coherence with "established" knowledge and practice'. Perhaps this is not a feature that matters much in the circumstances at hand. But that cannot just be assumed without good reason since it is so commonly expected in tasks like this. You will not be objective in designing the measure if you simply ignore the representational theory of measurement without good excuse (e.g. by not bothering to show why your proposed procedures will assign accurate values for inflation as characterised or defending that it will do so with dicey assumptions).

Measures of inflation also serve a number of external functions. They are used to update pensions, wages, and benefits, to understand effects on family budgets, to predict unemployment levels, to inform policies of the Federal Reserve and Central Banks, and so on. Which matter in the circumstances? Sometimes it is one, sometimes it is many all at once. And if it is many, what is the right balance among them? That is seldom made explicit. You have to figure it out. Objectivity requires due diligence at this.

We describe finding the complex of internal and external aims and finding the relevant ways to achieve them as separate steps, though they are not independent of each other. The aims you settle on affect the practices you undertake. Any set of practices you undertake will have other effects than just the outcomes you are aiming for. These effects can bring other external functions into play. Inner city old people without cars aren't necessarily a special concern in designing a measure of inflation. But as soon as the procedures chosen give heavy weight to suburban discount outlets that these people can't get to, their welfare can become an external function that needs proper attention.

Aims can be especially problematic for complex products that are expected to function properly in the natural world, even to the point of conflicting. This is well recognised in engineering, as Zachary Pritle (2020) notes in a US National Academy of Engineering issue on complex unifiable systems:

> Stress engineers see the loads on the plane and the vibration frequencies caused by flight, and their design goals may want materials that are high strength to stiffen the plane and survive loads during operations. But the heavier mass of those materials might make the job of the aerodynamicist harder, as she seeks to protect speed and to design an efficient profile of the plane.   (67)

This makes the duty of objectivity all the more demanding. To be truly objective you must nose out all the relevant aims and try to find some reasonable way of balancing or accommodating them all if you are objectively to endorse the product.

Supposing a—revisable—choice of aims then, what qualifies the practices undertaken to achieve that aim as objective? Given all that's known about means/ends reasoning, we propose that being objective requires that you have 'drawn attention to what matters' to achieve that aim. And that your claims about what matters to achieving those aims are sufficiently warranted.

Dealing objectively with the aim requires accountably giving due attention to what is reliable in the circumstances to achieve the aim. This involves 'drawing attention to what matters' along two dimensions in tandem. First, you need to consider what factors to consult from among the mass of facts, intervening factors, assumptions, practices, techniques, activities, products, and so on. Some of these may be unexpected or not factored in from the start. Second, you must choose the procedural constraints that will guide you.

With respect to the first dimension, the number and types of factors to take as relevant and reliable in order to achieve an aim are open ended, but they are constrained by the circumstances and the aim. You need to survey responsibly the range of possibilities and then be sure to have good reasons for what you choose.

With respect to the second dimension, the procedural constraints that are chosen (e.g. empirical adequacy, robustness) should be ones that support that the reasons settled on along the first dimension are good reasons.

So, our claim is that there is good reason to use the loose everyday concept of OAWKI in enjoining objectivity in typical scientific endeavours, like creating a measure, a model, a theory, a narrative, or a technological device. That's because you often do not have clear ideas in advance of what will be required. You want the scientists involved not just to be objective but to find out what being objective amounts to in the circumstances. But what does it amount to in the circumstances? This may not be entirely fixed but there will be things that are certainly required and others that are not.

As with duty of care, the answer may contain quite precise elements even though these could not have been specified beforehand: 'You should have directed your client's attention to the small print in paragraph 7.2' or 'An objective review of the evidence about badger culling requires consulting more than the RBCT results. It requires looking as well at genetic, epidemiological, ecological, and environmental studies.' What the final requirements to be found are is fixed by two elements working together: *the loose Ballung notion of objectivity*, with all that has been said and written about it (and partially recalled in Section 3.2.2 above), in interaction with the *context*. Context is supposed to help settle, within reasonable bounds, what you are to do to be objective.

This does not entail shying away from disagreement about either aims or methods. Consider what Worsdale and Wright (2021) argue in criticising the defenders of a particular measure of gender inequality: these defenders

> fail to offer a proper, contextualised assessment of the comparative objectivity of [their index versus a competing one]. Such an assessment would require attending to the context in which each of the indices is intended to offer reliable information, and evaluating whether the mitigation or elimination of certain perspectives in the construction of each index successfully diffuses the most significant threats to the reliability of that index's findings, relative to that intended context of use.
>
> (Worsdale and Wright 2021, 1675)

Worsdale and Wright conclude their paper with this advice:

> ...it is especially important for social scientists working...where the goals of research are subject to intense disagreement, to be clear on the contexts within which their claims to objectivity should be evaluated.
>
> (Worsdale and Wright 2021, 1681)

In their paper the two authors talk about the 'development policy' context as the one at stake in this dispute. But the problem is that that context can be characterised from a variety of different moral, political, and cultural perspectives and that will continue to be the case even if more description of the context is preset.

Often central features of the context are not in dispute. That is surely the case with our Vajont dam example of Section 3.4.2—nobody will challenge that safety of the surrounding villages was one of the aims that needed to be given great weight. But sometimes they are in dispute. That, we think, is the human condition—moral, cultural, and political differences abound. We have nothing to offer on how to negotiate them. It means that what is and is not objective is often a controverted matter, and that kind of controversy can spill over into the scientific domain as well, as illustrated in the case of the indices for gender inequality that Worsdale and Wright discuss.

We conclude from our discussion that whatever context is, it must be able to allow among relevant agents what we call (in the terminology of Andrea

Woody that we use in Section 4.5) a 'shared understanding' of both suitable aims and suitable ways to achieve them, enough to fix clear cases that are okay and clear ones that aren't, although much will be negotiable at the edges. Who, though, is a relevant agent? Injunctions to objectivity will mis-fire if there is no basis for reasonable argument, that is, there is not enough in the body of facts and information available so that those who are commissioned to be objective and those who are to pass judgement can produce reasonable arguments about what is done. This means that what is available in the context restricts the agents who can responsibly be involved on either side. Recall, not everybody will be up to shouldering a duty of care (e.g. to bank customers). They just can't catch on to what it is that they are supposed to do and, especially, to avoid doing. Similarly, given what is available in the context to set the bounds, not everyone will be up to the task of being object-ive at any given task in any given setting, where who might be and who might not will vary from case to case and, like the aims and practices, it may be vague just who can be in and who out.

Specifically, context must be able to do three jobs. Context must, first, pro-vide grounds for a good enough shared understanding of the purposes to be served in the circumstances, to then, second, settle well enough what prac-tices are relevant to achieving these. And, supposing as we will, that you must also be well enough justified in these choices, context must also, third, provide the materials for justification. We will use the example of the Vajont dam disaster to illustrate in the next section.

3.4.2 Objectivity in Context: The Vajont Dam Disaster

In early 1960s Italy, a large political and economic implementation plan was underway to transform an area of the Dolomites particularly rich in water into the biggest provider of electricity in northern Italy. The Vajont valley was part of this area, and works were undertaken to build the tallest (at the time) arch dam in the world there (Figure 3.2). The surrounding rocks in the Vajont area, however, had a type of structural fragility which, despite several premonitory signs and worrying conjectures about the his-tory of the place, was unfortunately overlooked. This resulted in an entire town being completely wiped away by a gigantic wave of water (with peaks over 200m) overflowing the dam on 9 October 1963. The overflow was due to a massive landslide (260 million cubic metres over an area of 2 sq. km) falling into the reservoir, and concomitantly, to the dam resisting the

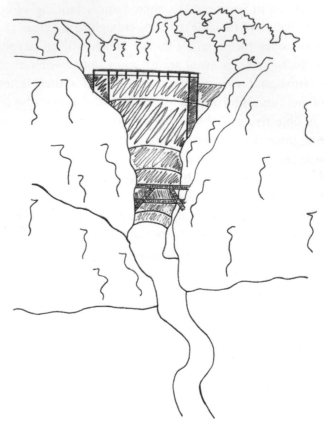

Figure 3.2 The Vajont dam
Source: Lucy Charlton.

impact of the slide, pushing the mass of water into the valley, and filling it in a matter of seconds.[32]

How could things go so wrong? How could the Adriatic Energy Corporation, the company in charge of the project, not take the risk of disaster more seriously? How could Carlo Semenza, the engineer who designed and chose the location for the dam, act so recklessly? These are some of the questions that haunted public opinion, the media, and the Italian political scene for years after the disaster. Science was attacked for being in cahoots with economic interests. Some of the scientists and experts stood trial,[33] and the charge was

[32] For a detailed description of the case, see Barrotta and Montuschi (2018a). Most of the literature on this episode is only available in Italian.
[33] Semenza died two years before the disaster occurred. Alberico Biadene took over the project and brought it to completion. He was tried several times and finally sentenced to six years in prison (three of which were commuted for health reasons). One of his closest collaborators, Mario Pancini, haunted by remorse, committed suicide.

not so lenient as 'lack of objectivity' on their part. Nevertheless, could this concept (or a suitable reformulation of it) help us understand better what happened in the Vajont disaster? To answer we need further background to the case.

Studies on the structural stability of the valley had been carried out in a very limited way in advance. And in fact, as unthinkable as this would be nowadays, in-depth assessments were not required by the juridical norms at the time. The engineers in charge of the project mostly relied on well-supported knowledge of the nature and behaviour of the rocks typical of that area (i.e. limestone) (Carloni 1995, 13ff.). 'From a geological point of view', Semenza wrote, 'the rocks [of the Veneto region] are generally very good... Overall, limestone is honest because it reveals any flaws on its surface' (quoted in Gervasoni 1969, 11). In-depth geological studies were therefore considered unnecessary because, by relying on scientifically assessed typical behaviour, the rocks of the area did not raise visible concern.

So, the principal aim of the project was to build a good dam. This was the inherent purpose to be achieved. Semenza and his fellow scientists had the required well-established knowledge to pursue such a project—in aid of increasing the electricity capacity of northern Italy, which was an explicit external function. Safety undoubtedly was a related aim, but it was addressed as a corollary controllable by the existing scientific knowledge. To guarantee that the surrounding natural environment would not catastrophically 'interfere' with the project, the expert team had just enough general knowledge about the typical geological behaviour of the rock constitution of the valley, and a number of premonitory signs (tremors, noises) that they chose initially, by and large, to neglect, and subsequently to handle inadequately, at least as we judge with hindsight.

By giving priority to one particular aim the scientists downplayed the purported *relevance* of several facts, intervening factors, and related methods, techniques, and procedures for following these up. The scientists

(1) put too much focus on the inherent purpose and on how it subserved the successful achievement of the external function of increasing the electricity capacity of northern Italy and
(2) treated this external function as clearly identified independently of the circumstances.

Together, these made them overlook the complex context in which the dam was to be built, which picks out the aims that needed to be attended to in building it.

This shows that the first job context is meant to help with: *settling on the aims for the case at hand.*

We now turn to a second job.

In pursuing a particular scientific project (e.g. designing an arch dam to be built in the Vajont valley) there is a wide range of ways to get it done. If you're told to act objectively in pursuing a project you must pick the ways that get it done in accord with the obligation laid on you (be objective). Designing a dam that stands (internal purpose) can have effects on further purposes that you might or might not be obliged to serve under your duty to be objective, such as increasing Italian electricity capacity and keeping people safe. Each further purpose might itself affect still other purposes (e.g. making the Adriatic Energy Corporation richer, buying building material from friends to favour them, scoring a political point to win over the electorate) that your duty to be objective might not or should not require you to fulfil.

Let's carry on with our story. The scientists' actions led to what was to prove a fatal mistake. What prompted such a mistake?

On one side, as mentioned, the use of local knowledge was not deemed crucial. Semenza intuitively trusted a view that the general knowledge he relied on behaved (to use an image from Cartwright 1999) like a vending machine: 'Feed in a description of the situation, and the theory will spit out a prediction of what happens' (184). Applied to the case of the Vajont: feed in that the rock in the Vajont is limestone with no significant surface signs of danger and our background theory 'Limestone wears its flaws on its face' predicts that the limestone is safe.

But, as we noted in Section 1.5, 'Limestone wears its flaws on its face' is a generic. Generics aren't reliable vending machines. You can always get a prediction from them but when you open the packet it might not be the right product inside. To be warranted in supposing that general knowledge about limestone holds in situ depends on a number of assumptions about the specific environment where those rocks live. Backing these up requires attention to knowledge coming from different sources, including knowledge coming from the inhabitants of the valley.[34] What makes general knowledge relevant to specific circumstances is not the logical structure of a theory, nor any isolated claims, no matter how well established they are. The idea of relevance that matters here has nothing to do with formal, logical relations. It is instead

---

[34] And this was part of the problem: by undervaluing knowledge coming from non-scientific sources the experts also undervalued, and overlooked, the more general problem of how relevant local facts are to make general knowledge itself relevant in context. Making the general relevant to the particular is not an automatic, deductive trick played by theories and scientific laws, even the best-established among them. See Barrotta and Montuschi (2018b).

material and contextual. Material because it has to do with content, and contextual because it can only be evaluated within circumstances and conditions of applications.

Besides, what this particular case tells us is that relevance is not only world-guided—guided by the facts—it is also action-guiding: it directs the actions undertaken on the basis of factual assumptions, and it directs what factual assumptions are to be considered. Making the facts matter in the circumstances entails seeing some facts as suitable means to achieving an aim. Building relevance is an action towards an aim.

This illustrates the second job that context is meant to help with: *adopting practices expected to be relevant to achieving the aims of the scientific endeavour for the case at hand.*

Aims are achieved not because of some lucky conjuncture of practices. Practices need to be relevant to the aim and, for objectivity, accountably so. This follows from two conditions: first, that you find what practices prove to be relevant to achieving the aim and second, that you are entitled to assume that they are.

The first condition is fulfilled by formulating, with the help of contextual facts, which will include circumstance-local information, an empirical assessment of what can work in the circumstances. The second condition is fulfilled by considering the constraints on the assumptions of relevance coming from well-supported knowledge, both general and local to the setting.

Satisfying these two conditions does not generally rely on pre-established and agreed-upon rules or norms. Relevance is ultimately an act of judgement, built by argument on the basis of the facts, the norms, and the assumptions made available by a context. Despite this element of subjectivity, it can still be objective, in the sense that you prioritise the features from the various lists of objectivity-making properties that best suit the purposes appropriate in that context and that guide you to activities that can serve that purpose.

In the case of the Vajont, in using general knowledge of limestone to inform a specific situation (the stones in the Vajont valley) not all knowledge about limestone is required—and this is not only a question of quantity. Some facts will matter more, or less, than others in assessing the behaviour of these stones in situ. In weighing facts against each other a situational range and arrangement of contextual factors plays a crucial part, well beyond what is secured by even the best-established theory or law taken at face value.

Effectively, what Semenza did in the circumstances was to abstain from making a specific judgement of relevance. But absence of judgement eventually led to flawed judgement. Semenza was aware that building the dam carried risks, and some of these were heavily ethically loaded since they could affect the welfare of thousands of people. But he assumed he could handle them by making his general knowledge 'what mattered' to minimising those risks.

In November 1960—when safety concerns had become apparent—Semenza started building a bypass. In case of a landslide being set in motion by working on the dam (a possibility that started being taken seriously, though admittedly controversially, by part of the scientific community), this would prevent the rising of upstream water, which could have endangered one of the towns built on the slopes of the valley. Shortly before his death he commissioned the construction of a model of the dam and of the whole reservoir—a highly innovative move at the time—which was meant to assist in understanding the effects of a landslide on the reservoir and in measuring the speed of its fall. In other words, Semenza thought of the risks he was taking as if they were primarily an epistemic problem, even though they had evident moral import.

The moral implications of the risks taken should have been specifically brought into account when evaluating the acceptability of the hypothesis that the Vajont limestone wears its flaws on its face. They 'mattered' in terms of assessing the relevance of the evidence accrued for that hypothesis—a point that Richard Rudner (1953) raised about hypothesis acceptance in the early 1950s:

> [I]n accepting a hypothesis the scientist must make the decision that the evidence is sufficiently strong or that the probability is sufficiently high to warrant the acceptance of the hypothesis. Obviously, our decision regarding the evidence and respecting how strong is "strong enough", is going to be a function of the importance, *in the typically ethical sense*, of making a mistake in accepting or rejecting the hypothesis.[35]   (2)

The lack of sufficient consideration of the human risk as part of the evidence on which to assess the acceptability of the assumption that the local limestone was not flawed and would stand steady weakened the grounds on which this assumption was ultimately accepted. Lack of sufficient consideration was a consequence of a not-well-identified dialectic between the internal and external purposes of the project: the safety of the downstream inhabitants should have been included as a far weightier factor bearing on the design of a dam that worked in the surroundings chosen for it. The use of

---

[35] The distinction foreshadowed here is between type I error (the error of accepting a hypothesis when it is false) and type II error (the error of rejecting a hypothesis when it is true), which we will not take up. It is widely discussed in the philosophy of science literature. See, among others, Douglas (2000).

local knowledge was relevant both epistemically and morally—it had consequences both on what to take as evidence and on what this evidence makes you responsible for. Failing to see its relevance was, equally, both an epistemic and moral mistake.

Now, back to the problem of finding objectivity. Given all that is understood about objectivity as we know it, the objectivity that is to be found is fixed by the context. We have seen something of how this happens in the case of the Vajont dam. The questions that should have been addressed were of the following type:

(1) What aims, both internal and external, should the project serve?
(2) What practices are relevant to achieving those aims in the circumstances?
(3) What enters a judgement about what are relevant aims and what practices are relevant to achieving them?

These questions are all 'contextual'. If context is to help provide answers then context needs to include

(a) Information about the wider cultural/social/political/economic setting and its norms and customs sufficient to constrain internal and external purposes.
(b) Facts about the setting that constrain what practices can achieve the purposes that should be served in that setting.
(c) An epistemic background that includes assumptions sufficient to support the relevance of the purposes and of practices to the purposes.[36]

These are the kinds of questions you are asking when you ask: *have you been objective in the circumstances*? Formulating them in terms of objectivity brings out the normative dimension that we have been making much of. This guides you towards judging not only what matters factually in the circumstances, but what you are responsible for in handling the complexities of the situation. The case of the Vajont dam is a good illustration of our basic thesis (from Section 3.3.4): 'If you have a duty to be objective here, you must find out—better, figure out—what it is that matters here in these very specific circumstances.'

---

[36] We should add that we do not commit to any particular metaphysics of contexts. Our strategy is to first figure out what 'context' has to do, then back read from that what it must consist of to do that. We leave open the question of how metaphysically to think of these ingredients (like norms).

We have argued that the duty to be objective can require of you both moral and epistemic considerations. There can be failures in both. The moral failure of judgements of relevance by Semenza and the rest of the team of dam designers exemplifies a lack of attention to the full set of possible or probable consequences of adopting one set of practices over another and privileging one set of aims over others.[37] Here the consequences could easily affect human safety, so they are clearly ethically loaded. In a different context the consequences might not involve such strong moral considerations. There is nothing specifically moral in a practice involving investigations to understand organic combustion (unless we put these in the context of, say, burning a pile of books as a symbolic gesture or lighting a campfire in a dry forest). Its aim—determining how combustion is achieved, by means of what substances, in what proportions, and so on—relates to a number of consequences and further purposes some of which might call for moral considerations (symbolically burning books) and others not so (making the balance of compounds comply with the Lavoisierian system of chemistry). The duty to be objective will have a more specific moral overtone in the first case, a more epistemic one in the second case. But, in any case, the overall need to make internal and external purposes meet stands.

3.5 Final Remarks

Objectivity will be 'found' by making the right judgements of relevance about aims, about practices that can achieve these aims, and about the assumptions that support these judgements, little of which can be done by the manual. There is no standard, preset methodology to deal with how you are to settle for certain aims nor for how you are to choose the practices that serve those aims. There are only contexts which, by means of the ingredients we describe, assist in these choices. OTBF sees 'being objective' as part and parcel of the process of settling on contextually relevant practices for achieving contextually relevant aims. The duty to be objective is a duty to pick and achieve the *relevant aims* using the *relevant* practices warranted by the *relevant* assumptions and to figure out what these are.

We do not intend relevance as yet another suggestion for a criterion of objectivity to add to the already long list, a further way to describe objectivity in

---

[37] Here we do not enter the discussion about foreseeable and unforeseeable consequences. As we think of consequences in the context of purposes, we assume that the consequences at stake are what a competent and responsible agent is able to envisage.

the abstract. Rather, we intend our stress on contextual relevance as a defence of the importance of the loose *Ballung* notion of OAWKI. Sometimes there is good reason to substitute a more specific, more refined concept of objectivity in science for the loose OAWKI. But equally there is often good reason to want the full action-guiding force of OAWKI, with the full range of duties it imposes. Of course, any particular precisification imposes a duty, but using *objectivity precisified* requires specifying these duties at the start. For a great many cases, no precisification is possible in advance, even when we know roughly what the context is. Finding objectivity requires a careful investigation of the context. Without a context there is no objectivity to be found, only objectivity to be shopped around for. And even with a context, finding can be hard work.

We are left with the question: does objectivity, pursued via the idea of context we articulate, promote reliability, or is at least a good symptom of it? There are two sides to an answer to this question: on one side, objectivity—OTBF—prompts you to recognise that, in the given circumstances, you must give appropriate weight to factors relevant to appropriately formulated aims; on the other side, objectivity prompts you towards discovering, in those circumstances, what *is* relevant and for what reasons. So when objectivity is found, it should indeed help ensure that the methods you choose are likely to achieve the aims settled on.

Here we have argued that the facts, norms, and expectations in a context provide the basis for judging what both the aims and methods for achieving them there are. But, as we argue in Chapter 4, it takes a great tangle of scientific work in the background to give meaning to these. The context for building the Vajont dam makes the safety of those down the valley a relevant external function and thereby makes local knowledge about limestone flaws relevant as well. But what constitutes a 'flaw in limestone', what constitutes 'safety down the valley'? Perhaps building tunnels to let sudden bursts of water escape is relevant to safety down the valley. But what constitutes the right kind of tunnel to serve this end and what constitutes exactly the kind of safety that would achieve it? Without a rich network of other scientific products—theories, measures, concepts, models, experiments, and the like—there is no answer. We may all feel you are talking sense when you call for investigations about whether the limestone is flawed, but without a vast tangle of scientific work in the offing, you are not. So yes, we agree that objectivity promotes reliability. But any meaningful demand for objectivity presupposes the tangle.

We turn now to an explicit discussion of what a tangle is supposed to be and what it can do in aid of reliability.

PART II
THE TANGLE OF SCIENCE

4

The Tangle

Preview

We began Part I of the book by emphasising that the usual suspects—the scientific method or methods, rigour, and objectivity—could not individually, nor in concert, secure reliability. Each, we argue, has its own weaknesses, and it is clear that these weaknesses are not ones that are compensated for by the others. The limited scope of rigour is not made up for by the addition of the scientific method or objectivity. Nor do rigour and the scientific method help solve the problems of determining context for objectivity. Apart or together, they simply aren't enough to do the job assigned to them. This is not to say that they cannot help with the project of reliability—they are valuable tools for evaluating the reliability of scientific products—but they do not seem enough on their own or together to secure the level of reliability we have come to expect in science. What more is there to call on?

We have already started to offer an answer. Throughout Part I we began to show how all three of the usual suspects presuppose the existence of a complex set of connections between the various scientific products they draw on to do their work—the eponymous tangle of this book. Virtuous tangles, we will now argue, are a critical missing tool, one that inescapably underpins the scientific endeavour for reliability, suggesting as it does not just that the right kind of pieces need to be drawn on but also that these pieces need to stand in the right kind of relations to one another.

In this chapter we want to underline that all the working parts of the sciences matter to scientific successes, all their products and creations, from theories and models to concepts and measures, studies and experiments, data collection, curation and coding, methods of inference, narratives and devices, technologies, designs and science-informed policies. Our focus, as you know, is on their reliability—can they do what they are supposed to?— and on what supports this. We argue that each of these different kinds of scientific products is more likely to be reliable when backed by the 'right kind' of network of other scientific creations, which we are calling a 'virtuous tangle'.

The Tangle of Science: Reliability Beyond Method, Rigour, and Objectivity. Nancy Cartwright, Jeremy Hardie, Eleonora Montuschi, Matthew Soleiman, and Ann C. Thresher, Oxford University Press. © Nancy Cartwright, Jeremy Hardie, Eleonora Montuschi, Matthew Soleiman, and Ann C. Thresher 2022. DOI: 10.1093/oso/9780198866343.003.0005

This is what we see the sciences doing when they are successful, across the board from gravitational wave theory to the design of measures for childhood nutrition. We have observed this in hundreds of cases ourselves over many years and it can readily be seen in countless histories of science, as we illustrate in Chapter 5 with our discussion of Hasok Chang's work on temperature measurement and Robert Bud's on the development of penicillin. Building these tangles is business as usual in the sciences.

We can go beyond piling up observation after observation though. In this chapter we offer three different arguments to support our claim that the right kind of tangle is a symptom of reliability and is conducive to it.

The first is that standard theories of confirmation presuppose tangles of the kind we advocate. If the job you want to do reliably is telling the truth, the usual ways of confirming that a claim does tell the truth depend on an appropriate tangle in the background. So, no tangle, no confirmation.

The second comes at it from the reverse direction. What makes for unreliability? Standard post hoc analyses of the failure of scientific endeavours generally find as part of the reason for failure that something from the tangle that should have been there was missing or faulty.

Third is an argument from constraint. The more numerous and well-defined are the other pieces that a scientific creation must work with to achieve a goal, the fewer options there are about what it must be like.

In Sections 4.1, 4.2, and 4.3 we march through these three arguments in defence of the tangle. In the final three sections we provide a fuller account of what we mean by a tangle, what should be in a tangle for it to merit the label 'virtuous', and why that could make for reliability.

4.1 In Defence of the Tangle: From Truth and Confirmation to the Tangle

In this section we hope to convince you about the need for a tangle. You may not love it, but you can't leave it. Suppose when it comes to science you are only interested in evaluating claims and theories and you only want to assess whether they are likely to be true: you are concerned with *confirmation*.[1]

---

[1] Even if your basic concern is more broadly with *reasons in favour of the truth* of scientific claims, these reasons will certainly include confirmation by evidence. We are happy to allow that reasons may go beyond this. For instance, a reason in favour of a theory in physics may be that it satisfies various abstract symmetry principles. It seems to us that interpreting these reasons and defending them *as* reasons will equally lead to the need for a broad tangle of other reliable scientific products. But it is

Since these are scientific claims, presumably *evidence* will play a big role in this assessment. If so, that already casts you into the need for the tangle. We are going to argue this by looking at simple versions of confirmation theories for illustration, but the same lessons hold equally for more sophisticated accounts.

We begin with the trivial observation that no interesting scientific claim H comes confirmed. Rather, something else confirms H. In the simplest accounts that we use for illustration, confirming H implies finding (a sufficient mix of the right kind of) evidence E that supports H, where support involves, as in the two famous schemas, either a hypothetico-deductive (HD) relation (as we discuss briefly in Section 4.1.1) or some basic probabilistic relation (Prob). As noted, we illustrate with the simplest versions of each, which use these two central notions:

HD: $H \rightarrow E$

Prob: $\text{Prob}(H/E) = \text{Prob}(H) \times \text{Prob}(E/H)/\text{Prob}(E)$

'Evidence' is a highly loaded label. There is no such thing as evidence—there are just facts—until a vast body of scientific work turns these into evidence. In the end we suggest it is better to think not 'we have good reason to assume H because H is "confirmed" by the evidence' but rather 'our reason is that H is backed by a virtuous tangle of scientific work'. In trying to understand what supports a scientific hypothesis, fixing your attention on single bits like 'evidence' is like trying to see how the Jacana's eggs are propped up by focusing on the small rigid twigs among all the ingredients woven together to make the nest (as in Figure 4.1), rather than widening your vision to make out what it takes to make them supporting (as in Figure 4.2).

Figure 4.1 Sticks in a Jacana bird's nest
Source: Lucy Charlton.

sufficient for our point to show that confirmation does so since confirmation is so central to the defence of the truth of scientific claims. We also do not take a stand on what kinds of things E for evidence might include since we maintain that what we argue holds for a broad swathe of things normally adduced as 'evidence' in science.

Figure 4.2 The Jacana bird and its nest
Source: Lucy Charlton.

So, if you focus entirely on evidence, you get the wrong picture. It takes a vast network of scientific products properly entangled to make evidence in the first place.

Evidence is *constituted by* this tangle and can *only play its role in confirmation given the tangle*.

The overall plan for our argument for this claim is this. We start with HD and rehearse the well-known lines of reasoning that show the number and stretch of ingredients needed. Then we turn to simple probabilistic theories of confirmation to argue that both the choice of a probability measure to be used in them and the choice of the event space over which this measure ranges require a tangle of scientific work if they are to be warranted. What we aim to show is that you may start wanting to know just about confirmation, but you will eventually end with a tangle (as in Figure 4.3).

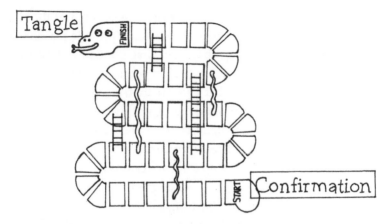

Figure 4.3 Starting with confirmation, ending with the tangle
Source: Lucy Charlton.

4.1.1 HD: Number and Stretch of Ingredients Needed

Here we argue for the need for a tangle from two notorious facts:

(a) HD is not enough
(b) $\neg (H \rightarrow E)$

4.1.1.1 HD Is Not Enough

As we describe it in Section 4.1, HD is not a method for deciding the accept-ability of a hypothesis but only a method for determining what can and can-not count as evidence for the hypothesis. It is widely acknowledged that just deducing a lot of facts from H is radically insufficient for confirming H. For confirmation you need more, what we call 'HD-with-bells-on'. HD-with-bells-on requires not only a set of Es that can count as evidence because deducible from H, but also that the Es that are implied are

  I. True
 II. Numerous
 III. Novel
 IV. Varied
  V. Differentiating between H and significant alternatives.

Nor is it enough for confirmation that there *be* enough of the right kind and variety of (true) facts implied. Science should provide reasons that support this. What, then, backs up the claim that, though not all the Es you adduce as evidence need be true, a sufficient number of the right kinds are? You need reasons for the claim that enough are true, that enough novel ones are true, that the set of true ones has enough of the right kinds of variation, and that they genuinely differentiate H from alternatives. We will illustrate with the last two requirements. These immediately expand the work that needs to be done well beyond just piling up evidence claims that there is good reason to suppose are true. Here already is the need for a rich tangle of supports.

Differentiating clearly gets you far from merely piling up evidence claims. To provide reasons that a set of Es, say $\varepsilon = \{E_1, E_2, \ldots E_i, \ldots E_N\}$, does this, science has to have a list of probable alternatives and also good reasons that it is a good list. Then, reasons are needed to support that this set rules for H and against these alternatives. But what supports the claim that the alternatives imply the negation of the evidence claims in the targeted set? After all, it is well rehearsed that generally neither H itself nor any of its alternatives imply much by themselves. They need auxiliaries. So what the alternative hypotheses imply depends on what auxiliaries hold. That means science needs to uncover and provide reasons for these auxiliaries...a big multifaceted task indeed. Since we discuss this in more detail in Section 4.1.1.2, let us turn to variation.

What kind of *variation* is needed and what reasons can be adduced in favour of, first, the claim that that is the right kind of variation, and second, that a large enough set of true E_is exhibit this kind of variation? As usual, the answer differs case by case. Let us look at one in particular involving the Bangladesh Integrated Nutrition Programme (BINP), which was proposed by the World Bank and initiated in 1995. We oversimplify a bit and also will stretch the evidence base beyond what was actually available, in order to highlight our point.

The hypothesis in question is this:

> In places like Bangladesh where 'behaviours related to feeding of young children have at least as much (if not more) to do with the serious problem of malnutrition...as poverty and the resultant household food insecurity do', the Integrated Nutrition Programme (INP) will improve infant nutrition.
>
> (World Bank 1995, 4, cited in White and Masset 2007, 630)

The main activity of INP is to educate mothers about how to use resources available to provide better nutrition for their children. This hypothesis turns out not to be true despite good reasons in its favour: it failed in Bangladesh.

As Howard White and Eduard Masset (2007) report in their post hoc evaluation, the basic theory is simple: where 'bad practices' are responsible for malnutrition 'changing bad practice to good will bring about nutritional improvements' (630). There were a number of good reasons in favour of this hypothesis before the Bangladesh failure was observed, going even beyond HD-type evidence, including, for example, both a general theory and a model of how the effect is supposed to be produced. The causal model described by White and Masset looks like this:

(1) The right people (those making decisions regarding [the feeding of] under-nourished children) are targeted with nutritional messages.
(2) These people participate in project activities, and so are exposed to these messages.
(3) Exposure leads to acquisition of the desired knowledge.
(4) Acquisition of the knowledge leads to its adoption (i.e. a change in practice).
(5) The new practices make a substantial impact on nutritional outcomes. (630–1)

The HD evidence came from Tamil Nadu where INP was introduced in 173 out of 373 rural blocks in eleven districts, and the evidence was positive.[2] The World Bank reports it as one of the 'most successful [projects] in the world in reducing malnutrition'.[3] Let us imagine for the sake of argument that it was trialled successfully in a number of other places as well.

The issue we want to raise is one about what constitutes '*the right kind* of variation'. In this case, post hoc there is a far clearer idea of the answer than was available at the time. The point we want to make is about how much science, and of how many different kinds, goes into discovering and supporting these claims about variation.

What went wrong in Bangladesh? Short answer: *men* and *mothers-in-law*:

The program targeted the mothers of young children. But mothers are frequently not the decision makers, and rarely the sole decision makers, with respect to the health and nutrition of their children...women do not go to market in rural Bangladesh; it is men who do the shopping. And for women in joint households

---

[2] See https://extranet.who.int/nutrition/gina/en/node/23345.
[3] See http://www.worldbank.org/ourdream/india_2.htm.

– meaning they live with their mother-in-law – as a sizable minority do, the mother-in-law heads the women's domain.   (White 2009, 6)

So, in order to be appropriately varied, the evidence base should have included considerations about places where the mothers live in households where they have little control over what is bought and little control over who gets to eat what.

What would support this particular demand on variation? What does it take to say, 'there is good reason to think the evidence needs this particular kind of variation'? Here is some of the scientific work that went into supporting post hoc the variation that would be needed for the INP hypothesis to be well tested. Notice how detailed and varied it is:

(1) Data from the Demographic and Health Survey, which 'show that only one in five women [in the BINP population] are solely responsible for decisions regarding their own health and that of their children, falling to only 1 in 10 of women living with their mother-in-law' (White and Masset 2007, 633).

(2) Propensity score matching: this was used to construct comparison groups.

(3) Focus group meetings and reports:
  - '...from these came the memorable quote, "we'll start doing that once our mothers-in-law have passed" ' (White 2009, 6).

(4) Semi-structured interviews with 150 individuals (White and Masset 2007, 631).

(5) Ethnography:
  - for example, White reports that Sarah White's book, *Arguing with the Crocodile. Gender and Clash in Bangladesh*, played a significant role (White 1992).

(6) Statistical techniques and econometric modelling (as in Figures 4.4 and 4.5).

4.1.1.2 $\neg(H \rightarrow E)$

Now couple this with the widely stressed reminder that seldom does H imply a confirming fact on its own. Rather, it takes a number of *auxiliary assumptions* to join with H to imply E_is of interest. Note that auxiliaries will generally be required for each of the E_i adduced in the body of evidence for H. So you need

$$H \,\&\, A_i \rightarrow E_i, \text{ for EACH } E_i, \text{ where each } A_i \text{ is generally a big set } \{A_{i1} \; A_{i2} \ldots A_{iN}\}$$

Table 2. Women's say in decision-making by position in household (percentages)

	Own health	Child's health	Daily purchases	Cooking
All respondents (n = 9716)				
Respondent	20.6	19.2	19.5	66.8
Joint	36.1	47.1	43.8	20.4
Husband	36.9	26.9	27.2	4.6
Someone else	6.4	6.9	9.6	8.2
Currently married women in male-headed households (n = 8706)				
Respondent	15.7	14.3	15.5	62.8
Joint	36.8	50.0	46.1	19.3
Husband	41.0	29.9	30.2	8.7
Someone else	5.1	5.8	8.2	9.2
Living with mother in law (n = 2017)				
Respondent	12.2	10.3	9.8	42.8
Joint	36.2	48.5	45.2	31.3
Husband	39.5	27.1	22.8	3.3
Someone else	12.2	14.5	22.2	22.6

Note: The sample for DHS is ever-married women aged 10–55.

Source: Calculated from 1999/2000 DHS data.

Figure 4.4 Data on women's say in decision-making in Bangladesh
Source: White and Masset (2007), 634.

Taking into account what this involves as each of the requirements I.–V. is satisfied, you end up with a very taxing set of demands.

'H is confirmed' implies that there is a set $\{<A_1, E_1>, <A_2, E_2>, \ldots, <A_N, E_N>\}$ such that

- Each A_{ij} and E_i are backed by good reasons
- The set of pairs is large
- The set of pairs is varied and in the 'right' ways
- Enough of the E_i are novel
- For each E_i, there are good reasons against explanations alternative to H&A_i.

It takes a vast tangle of scientific work to satisfy these requirements. So, looking for confirmation casts you right into the middle of the need for the tangle.

None of this is news of course. It is patently clear from the HD form itself coupled with the failure of *the given* or any other secure basis on which to found knowledge. The full significance is easily obscured, however, by calling auxiliaries 'initial conditions', or 'boundary conditions', thereby suggesting

Table 5. Knowledge and KP gap equations

	Rest during pregnancy		Colostrum		Breast-feeding	
	Coefficient	t-statistic	Coefficient	t-statistic	Coefficient	t-statistic
KP gap equation						
Daughter-in-law	−0.04	−0.40	−0.07	−0.68	0.30	2.30*
H/h owns land	0.00	1.63	0.00	−0.07	0.00	0.95
Farming household	0.14	3.44***	−0.07	−1.67*	0.12	1.99**
Durable index	0.00	−0.16	−0.01	−0.77	−0.10	−3.46***
No. of children	0.09	4.58***	−0.06	−2.69***	0.02	−0.53
Mother's education	−0.01	−0.30	0.03	1.01	0.04	1.38
Father's education	−0.04	−1.35	0.03	1.00	−0.04	−1.36
Elderly male in household	0.11	1.90*	0.03	0.60	0.03	0.35
Project area	0.13	1.25	0.03	0.26	−0.13	−0.58
Project*durables	−0.01	−0.72	−0.01	−0.69	0.07	2.49**
Project*daughter-in-law	−0.06	−0.57	−0.01	−0.09	−0.28	−1.81*
Project*children	−0.03	−1.28	0.03	1.00	−0.01	−0.22
Intercept	−0.95	−11.07***	−1.08	−10.17***	1.57	5.66***
Selection term	1.44	5.82***	1.89	3.33***	0.61	2.74***
Observations	5391		4266		5395	

*Significant at greater than 10%, 5% and 1% levels, respectively.
**Significant at greater than 10%, 5% and 1% levels, respectively.
***Significant at greater than 10%, 5% and 1% levels, respectively.

Figure 4.5 Knowledge gaps in participating mothers in BINP

Source: White and Masset (2007), 639 (boxing added).

they are easy to secure. C.G. Hempel (1988) expressed similar worries about the vague 'ceteris paribus clause' often attached to scientific principles. 'It is too readily used to vouchsafe the deductive potency of a theory when no such thing has been achieved' (156–7). We could actually argue the reverse: it is the non-deductive potency of the auxiliaries (the vast tangled work required to secure these and justify including them to support confirmations) that cast doubt on mechanically (deductively) achieved confirmation. (For a discussion concerning some of these points, see Barrotta and Montuschi 2018.)

Turn now from confirmation by HD-with-bells-on to some brief remarks on probabilistic confirmation.

4.1.2 Probabilistic Theories

There are a great variety of probabilistic theories, with most admitting both objective and subjective concepts of probability. Most employ either Prob (H/E) or Prob (E/H). In any case, a probabilistic theory requires

- a probability measure Prob (...)
- Over an event space that contains E and H and presumably a great deal more: (..., E, ..., H, ...).

Where does this probability measure come from? If it is subjective, the usual answer is that it comes from, or at least is constrained by, 'background' knowledge. But if the background knowledge is to be good enough to set a probability measure that you are warranted in using for confirmation, it will inevitably involve a great tangle of solid scientific work. If it is objective—that is, there really is a true probability to be found—science will equally need to include a great tangle of work in support of the estimate made of it.

The same thing is true with respect to the choice of event space. What are the *right* features to include in an event space that is to be used to evaluate how well H is confirmed? And what about those features themselves? Good science requires that the concepts employed for those features have solid backing, and that, as we have stressed, involves both the development of measures for all of them (which we discuss again in Section 4.2) and legitimation, not just that they are good concepts that apply in the world in general, but specifically for the purpose at hand.

Think again about the Integrated Nutrition Programme, where success in Tamil Nadu was taken to provide positive confirmation for the hypothesis that in places where the behaviour related to feeding of young children plays a significant role in malnutrition, educating mothers in tandem with other aspects of INP will improve infant nutrition. The post hoc analysis showed that the evidence from the Tamil Nadu study was expressed in the wrong concepts. It was not 'educating mothers' that produced good outcomes but rather a feature that, in the Tamil Nadu context, was co-extensional with 'educating mothers', to wit, educating those who

- control what food is procured
- control how food gets dispensed
- hold the child's interests as paramount.

We call this the *Donald Davidson problem*. Davidson (1995) maintains that causal claims can always be captured under a causal generalisation. If an intervention x causes an outcome y in some context, then there are descriptions of x and y, say C and E, such that it is a general truth that Cs cause Es.[4] But, Davidson warns, the concepts that appear in the universal generalisation may not be the ones we use to describe the intervention and the outcome. He urges, for example, that the events on the cover of *The New York Times* may well have caused the events described in the continuation of that article on page 6—but not under those descriptions. Solving the Donald Davidson problem to settle on the right concepts to use in an event space appropriate for evaluating degrees of conformation is clearly no easy matter. Again, a large tangle of work is going to be required to support that the event space chosen is up to the job.

4.1.3 In Sum

Although in this section we have looked at the skeletons of confirmation theories and not dealt with real, live, fleshed-out theories, the kinds of arguments we give carry over in spades to more sophisticated versions. The lesson is that you can't do confirmation and have good reasons for taking your confirmation to be reliable without a virtuous tangle to back this up.

---

[4] Presumably he would amend this to allow the kind of concatenation of causes we discuss in Section 2.7: if there is no such C and E there should instead be a series of intermediate happenings between x and y for which there are descriptions that bring each cause–effect pair under a general causal principle.

4.2 In Defence of the Tangle: What's Happened When Things Have Gone Wrong?

What do you do in science when a scientific product does not do what is aimed for and you want to figure out why? You look for what pieces are missing from the tangle to back up the product that you think should have been there in the first place and for what was presumed that shouldn't have been. You may also adjust the aims that you expect the product to serve, which we discuss further in Section 4.5.

So far, we have mentioned the following kinds of items that could be missing:

- If you expect a claim to be reliable for providing an accurate or true report then, as we note in Section 1.5 with the Post Conviction Risk Assessment (PCRA) algorithm, you should be worried if various types of validation are missing for concepts in that claim, like construct validity, content validity, convergence validity, and so on.
- From Section 1.5, if you want a measure you have designed to be reliable for measuring a concept as you characterise that concept, you should be worried if good arguments are not available (or even better, formal representation theorems) that
  - (a) the representation provided in the measure is appropriate to the characterisation
  - (b) the procedures will assign accurate values of the concept to the systems that are to be measured.
- If you want a particular concept (like *annual level of inflation*) to be reliable for a specific purpose (like setting benefits for veterans) you should be worried if there are no arguments available that the particular precisification of this concept to be used for making measurements of it (like CPI) can do the job required (as in the example from Section 1.5, about adjusting veterans' benefits to keep veterans' standards of living stable from year to year).
- In archaeology, as we illustrate in Section 4.4, if you want observations that can reliably serve as evidence (both sound and relevant) about what a site was for (e.g. was it a burial site?) you should be worried if there is no good—sound and well-warranted—'middle-level theory' available to interpret the material items found in the site into the more abstract terms in which archaeological evidence is expressed.
- If you expect a claim of the form 'Generally X' to reliably predict that X will hold in a specific setting D, you should be worried if knowledge

relevant to whether D is likely to be typical or atypical with respect to X is not part of the backup (as in the Vajont dam disaster from Section 3.4.2).

Filling gaps in the tangles that support that scientific products can do what is expected of them, eliminating what are thought to be faulty pieces, and honing the aims to better fit what the product can do is 'business as usual' in science and in science-based policy, engineering, and technology.[5] This is part of the point of demanding that these are critical community enterprises. What you see in the bullet list above is absolutely typical across scientific disciplines and science-informed technology and planning. It is essential to good science to look for the gaps and failures that might undermine success and find ways to fix them.

Here are a few further examples, across a range of scientific contexts:

- In Robert Millikan and Harry Fletcher's famous experiment to measure the charge of the electron, a charged oil drop hovers between two plates, pulled down by gravity and up by the electromagnetic field created by the potential difference between the plates. It was essential to know the potential difference precisely. So Millikan and Fletcher did not just trust to the reliability of the battery voltage. They also measured the potential difference in six parts by a device accurate to 1 part in 2,000. This device in turn was calibrated by a second whose accuracy was certified and also independently measured by yet a third device. And 5,000 readings calibrated in two different ways were shown to be consistent.

- The gyroscopes that the Stanford Gravity Probe put into space to test the General Theory of Relativity were precisely designed so that all other causes of procession than coupling with space-time curvature were eliminated. Nevertheless, just to be sure, the spaceship was to be rolled to distribute other possible effects evenly.

- Or consider meta-analysis of experimental results in medical research. As one advocate (Haidich 2010) puts it, 'A failure to identify the majority of existing studies can lead to erroneous conclusions' (29). But ways have been developed to plug this gap, as the author notes: '…there are methods of examining data to identify the potential for studies to be missing; for example, by the use of funnel plots' (29).

---

[5] What these are taken to be will, of course, depend on what scientific perspective is assumed, and there can—and often are—debates about this between perspectives. We discuss this further in Section 4.6.

- Another example is the work at CEDIL (the Centre of Excellence for Development, Impact and Learning), which was sponsored by the UK Department for International Development, to identify and plug where there are gaps in the methods available to evaluate the effectiveness of development programmes. The authors of the CEDIL report (Davey et al. 2017) on this explain: 'We conducted an interdisciplinary consultation and literature search to identify "gaps" in methods used to evaluate interventions in international development. We differentiated between gaps in the adoption of existing methods and best practices, and areas where methodological research is needed .... "Gaps in use" are gaps between expert consensus and use of the methods. Methodological innovation is not needed to close these gaps; the challenge is awareness and implementation. "Gaps in methods" are where innovation is needed to improve learning from evaluations. This report is focused on the latter gaps...' (4–5).

4.3 In Defence of the Tangle: Constraints Make It Hard to Misstep

Section 4.2 offers evidence in favour of the need for a tangle by pointing out that when things go wrong, post hoc diagnoses regularly point to a missing piece that should have been in the tangle if reliability was to be expected, or a piece that was there but flawed. Here we offer a simple mechanism by which a good tangle promotes reliability, though we have far more to say about how well this mechanism can work in Section 4.6 and in the Afterword. The root idea is simple, coming from reflection on how things go wrong: a tangle makes failings less likely by supplying a network of constraints that makes it hard for something to go wrong.[6]

We build on the argument that Sharon Crasnow (2021) offers in support of her notion of *coherence objectivity*.[7] Though she does not use our language, coherence objectivity calls for just the kind of tangle of scientific work that we defend. We will follow Crasnow in using as an example the V-Dem measure of democracy, which will connect to topics in Chapter 6. Among the

---

[6] We are not the first scholars to portray constraints as a productive force. Caporael et al. (2014), for example, have examined scaffolding during development, arguing that it 'implies generative entrenchment: differential change or transformation productively constrained by earlier development' (16).

[7] Crasnow also talks extensively about 'reliability' in her paper. But she is concerned with various specific senses of this term used in the social sciences, not with ours that a scientific product is reliable for a certain purpose if, when it is used as planned with the other pieces planned, the purpose will most likely be achieved.

sense of objectivity we describe in Section 3.2.2, the ones that Crasnow engages with are 'those having to do with representing the object of inquiry accurately...and intersubjectively'. We can recast this enterprise in our terms as an attempt to explain *when V-Dem indices are reliable for accurately determining the degree of democracy in states they are applied to.*

But beware: in Section 1.5 we warn of purposes that are too general and can lead to inference by false ascent from the concrete to the abstract. This is clearly one of those, which Crasnow, too, acknowledges in her arguments that coherence objectivity is inextricably value-laden. Different purposes require that different values be placed on different constituents identified for democracy and for their constituents in turn. This gets reflected in different weightings in constructing a single overall index either for democracy or for its constituents and their sub-constituents and hence results in different indices. (V-Dem provides separate indices for five different constituents, or what V-Dem calls 'types' of democracy, but does not amalgamate them into an overall index.) To avoid too much complication, we will nevertheless suppose for the sake of discussion that this purpose is well-enough posed.

Crasnow argues:

> It is a coherence of measures *with theory, empirical knowledge, and practical knowledge* that provides support for assessing the validity of the measures.
>
> (1224, emphasis added)

and

> [D]emocracy might be understood as measured objectively when it is *subject to theoretical, empirical, and normative constraints.*   (1207, emphasis added)

She then describes the kind of theoretical, empirical and normative work involved. In our framework, these help make up the tangle that supports the reliability of the V-Dem indices for measuring democracy accurately. Here are a just few examples from the ones she talks about:

- Empirical research investigating 'transitions to democracy, backsliding from democracy, the relationship between democracy and peace (the Democratic Peace literature), democracy and economic development, democracy and human rights, and others' (1207, emphasis added).
- The complex of theoretical and empirical work that filters out five constituent features central to the understanding of the *Ballung* notion of democracy (e.g. electoral democracy) and that also locates subcomponents (e.g. freedom of association and freedom of expression).

- A variety of coding methods (i.e. methods for how to provide numbers for the variables measured).
- Evidence of intercoder reliability.
- Studies to show consistency with other indices.

Following Hasok Chang, Crasnow stresses the need for these theoretical, normative, and empirical constraints to *cohere*. What is meant by 'coherence'? Recall from the Preface that Hasok Chang has been making great efforts to explain that. We don't have anything useful to add on that general question. When it comes to our issue of reliability we think that it is a case-by-case matter both what kinds of elements are required in a tangle and how they relate to each other.

What we do have something to say about is the job that the elements of the tangle do, taken together, when they make for reliability. We see Crasnow's work on objectivity as providing part of the answer. In terms of Crasnow's example, what is the tangle doing that supports the reliability of V-Dem to measure democracy accurately? Crasnow's answer: 'constraint'. This term appears ten times in her paper. Crasnow supposes that among the factors she cites there are enough of the 'right' ones that *they together constrain the V-Dem indices and the targeted notion of democracy sufficiently so that these indices are likely to measure democracy in the sense intended accurately.*[8]

The factors Crasnow cites are just the kind that we discuss in Section 4.2, ones that you would expect to turn up in a post hoc diagnosis if the constructed measure goes wrong. But notice, if the V-Dem measure-constructed subject to these interlocking constraints is to be reliable in measuring democracy, these in turn must be reliable: they must be reliable for the role they play in constraining V-Dem to measure accurately the concept it is supposed to. This is why support for the reliability of any particular scientific output involves an expanding network of support for the reliability of a great number and variety of other scientific endeavours.

Let's explore how this works. As noted, V-Dem measures are constrained by empirical and theoretical research, which includes research on the causes, effects, and correlates of democracy. For instance, besides the theory of the democratic peace that we look at in detail in Chapter 6, there is widespread agreement that democracy is correlated with levels of education and with

---

[8] It is worth noting again the caution we raise about loosely specified purposes. For the loose *Ballung* notion of democracy, what counts as accurate measurement? One uncontroversial constraint is that the measure should get right both the cases that are clearly in and those that are clearly out. Standard exemplars of these that Crasnow (following V-Dem) notes are Sweden and North Korea. V-Dem gets them right.

levels of income. (But there is dispute about which way the causal arrows run and in some studies a common cause is suggested.) This means that those states that V-Dem counts as democracies should have high levels of education and income per capita, or, more carefully, that if they do not, a good reason should be at hand. This, though, is only required if V-Dem is setting out to measure democracy in the same sense as used in the studies arguing for the correlation.

What notion of democracy, then, is V-Dem setting out to measure? Rather than offering a precise characterisation, V-Dem aims to maintain a multifaceted *Ballung* notion, which is probably part of why Crasnow chooses to study it. Here is what V-Dem's own comparison with other measures (Coppedge et al. 2017) says about this:

> While most other projects are focused on developing one or two very high-level indices, V-Dem is focused on the construction of a wide-ranging database consisting of a series of measures of varying ideas of what democracy is or ought to be . . . (19)

V-Dem does so because

> It is important to recognize that the core values enshrined in the varying principles of democracy sometimes conflict with one another. Such contradictions are implicit in democracy's multidimensional character. Having separate indices that represent these different facets of democracy will make it possible for policy-makers and academics to examine potential trade-offs empirically. (24)

This sounds as if it will be hard to tell whether V-Dem measures the same concept as the different measures used in the various correlational studies. On the other hand, as a recent comparative assessment of measures (Boese 2019) reports,

> The three ['most prominent'] indices [V-Dem, Polity2, and Freedom House] display a high level of agreement for those observations included in all three datasets. (95)

The kinds of consistency-with-other-measures validity studies that back up this claim are an almost essential ingredient in a supporting tangle for any measure: be wary if they are missing. But also note that they are not enough, as we discuss in more detail in Section 4.5 when we expand on what a virtuous tangle should consist in.

Controversies over the causal explanation of observed correlations are another huge source of constraints: adversaries are quick to point out possible flaws in the measures used in opposition studies. We won't pursue this here, however, since we look at an analogous source of constraints in detail in Chapter 6 in discussing democratic peace theory.

The vast tangle of scientific work in the background simultaneously constrains both a measure (or model, theory, technique, etc.) and the aims it can be expected to serve. This means that there is not much wiggle room for what these are like. The start of this section claims that this makes mistakes difficult. Still, we do not want to maintain that reliability is ensured. The best science can do is to produce enough constraints that reliability is to be expected from the point of view of the science that does the expecting. But science, as we know, doesn't always get it right. A vast number of ways of being mistaken are ruled out but a vast number may still remain.

That there are no guarantees should be no surprise, at least for those who reject any kind of foundations for knowledge. Reliability sits alongside truth in Neurath's boat. According to Neurath ([1921] 1973),

> We are like sailors who on the open sea must reconstruct their ship but are never able to start afresh from the bottom. Where a beam is taken away a new one must at once be put there, and for this the rest of the ship is used as support. In this way, by using the old beams and driftwood the ship can be shaped entirely anew, but only by gradual reconstruction.   (199)

Nevertheless, the existence of a virtuous tangle is a strong indicator of reliability because of the thick network of constraints it provides. And we are in a better position with reliability than truth because reliability is a weaker demand than truth. If you want a claim, or model, or measure, or technique to do the job you put it to, for this the claim need not be true,[9] the model need not provide a true representation of what is modelled, the measure or technique need not in fact be what you think it is or work the way you think it does. This does not disbar them from succeeding at the work you want from them.

That this is true is clear from the history of science, as is widely acknowledged. Many, many sciences that are now abandoned as false (or at least not entirely true) were nevertheless successful at a vast array of uses to which they were put. But—at least from the point of view of more modern science—they could not mean what the scientists of the time thought they did, since,

---

[9] Unless that is the specific job we want to get done.

by the hypotheses of modern science, many of the concepts these older theories used do not refer to anything that obtains in the real world.

Recall our discussion of meaning as use in Section 1.5, **Argument (i)** a, of generics in **Argument (i)b**, and of our claim in Chapter 2 that it takes a network of other scientific endeavours to turn facts into evidence, repeated in Section 4.1. For most facts that play a role in science, what the fact implies—what must be the case if it is to genuinely hold—depends on what other facts there are. This is frequently true even for general facts since a great many general facts are expressed in the form of generics, many of which refer to dispositions that may not always be present, may need triggering, and may be readily overwhelmed or cancelled out by competing ones. For instance, whether 'people respond to incentives' implies that you here now will respond to a particular incentive depends on other facts like how habit-bound you are in your current activity, how counter-suggestable you are, whether you recognise what is on offer as an incentive in the first place, and many more. This is mirrored in language by the need to establish meaning through use. Together, these considerations paint a gloomy picture about science's prospects for understanding things as they really are since it means that large networks of use must match large swathes of facts. Still, as we have noted from the start, science produces a vast array of successful products. Science's trial-and-error efforts to create ever more products to put in the scientific warehouse (certified as reliable for specific jobs) and to produce ever more virtuous tangles, can help explain how this can be possible.

Suppose, for instance, that you build a model for a helium-neon laser and predict from the model that devices built according to it will emit a coherent beam at 633nm. And they do. There are hundreds of products—concepts, models, experiments, and more—that cooperate together in and around the model to allow it to produce this prediction reliably. Each of these products plays a role, in tandem with a whole new set of cooperating partners, in certifying the reliability of other successful products that do other things. And each of the products in each of the new groupings, like each of the originals, participates in myriad other successful products. And so on.

Suppose now that some of the assumptions that underwrote your confidence in the reliability of the helium-neon laser model you started from are mistaken—the products you took to underwrite the model's reliability are not as you think. The concepts you use to describe these products or your characterisation of what these products are reliable for may not match features in the world as they really are, or the many additional assumptions you suppose, albeit many implicitly, in fixing the content of your first set of assumptions do not correspond to real facts, and this again either because the concepts these further assumptions employ are amiss or the assumptions that give them content are mistaken, and so on. Nevertheless, the laser did emit the 633nm beam that the model predicts.

We suppose, at minimum, that nature is not entirely capricious: there are consistent, intelligible, and systematic patterns even though science may never recognise correctly what they are. This seems required if successful prediction based on the accumulation of reliable scientific products, know-ledge claims, and practice is to be possible, as we know it is. It is not just random chance that lasers built according to this model, whose reliability is backed as it is, emit the beams predicted. They do so because of the features the parts actually have when assembled together and the principles that actually govern them in that arrangement, even if these are not as science supposes them to be. And, with regard to the network of products that you take to certify the model's reliability, unless it is pure chance, that network must succeed in picking out features that, governed as they truly are, do in fact produce a coherent beam.

Recall that we are assuming that each of the products you took to certify the reliability of your model appears in successful products designed for other aims. Being realistic to what really happens in science, we suppose as well that they were tried out, with something like the original scientific understanding and also modified understandings, in many products that failed. In all these cases we suppose the products succeed or fail because of what features they actually have. When the scientific system is working well, the understanding employed in another new product will be constrained by what science has learned of what it fits with and what not across the panoply of successful and unsuccessful scientific creations. Throughout its endeav-ours, science can only make use of its own principles and its own image of what the world is like. We do not want to assume that success implies that science is getting these right. Still, success is often achieved. So, science may not be learning about the true features and the principles that govern its products, but it does learn how to confine its uses to those that rely on true features.

Think of this analogy between this kind of scientific tangle and the scien-tists engaged in it and Meccano and Meccanomen. (Meccano is a construc-tion system for serious builders—known as Meccanomen, independent of gender.) You have in front of you a large collection of a great variety of Meccano parts. A typical collection could be expected to have many thou-sands of parts of more than six hundred different kinds. Each part has a number and a label: '5: perforated strip 5 hole', '10: fishplate', '27a: spur gear 57 teeth', '224: flexible triangular plate 7 × 3' (Figures 4.6–4.9).

Figure 4.6 Meccano 5 hole strip
Source: Lucy Charlton.

Figure 4.7 Meccano fish plate
Source: Lucy Charlton.

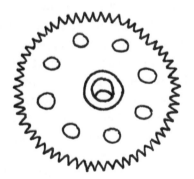

Figure 4.8 Meccano spur gear
Source: Lucy Charlton.

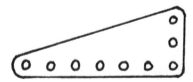

Figure 4.9 Meccano triangular plate
Source: Lucy Charlton.

Suppose you have also recruited a number of Meccano builders, each with different interests. Some like designing buildings, others transport devices (trains, planes, cars), others design work tools (tractors, bulldozers). Each of the builders has access to this common pool but also to a separate set of different kinds of parts just for themselves that they use in their own designs.

As we noted, one of the important lessons of the history of past successes in science by theories and practices that science now thinks could never have been true, is that things can work for reasons other than those that were supposed in their use. That is what we imagine in our tale of the helium-neon laser model. Similarly, a causal model in the social sciences may make very good predictions across the range of domains you apply it to yet not, contrary to what you suppose, picture correctly the dominant causes producing the outcomes the model predicts. And so forth.

We want our analogy to reflect this. So we shall suppose that our Meccanomen never actually get to examine the pieces or even see them, so they don't know what they are actually like. Long ago, after their first simple Meccano sets, they stopped physically putting the parts together themselves. Since then they have only provided designs for others to construct. So they have no immediate knowledge of many of the parts, though each has a conception of what a part with a given label is like. Let's suppose that there is a reasonable overlap in the characteristics they attribute to the parts but also some differences, as with the different conceptions of democracy underlying different measures.[10] Suppose that, for your collection, many of the labels are wrong. Things are not entirely as your Meccanomen take them to be. For instance, the part labelled '5: perforated strip 5 hole' really has only three holes in the middle plus an indentation on both ends (Figure 4.10) and '27a: gear wheel 57 teeth', which is supposed to have a number of holes within its toothed circumference, has no holes at all (Figure 4.11).

Suppose these builders have been doing these hands-off designs in their area of interest for a good while and they watch what each other are doing, the successes and failures and the diagnoses of what parts of the plan have gone wrong when failures occur. At first, most were probably not very

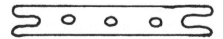

Figure 4.10 Meccano impostor 5 hole strip
Source: Lucy Charlton.

---

[10] As a matter of fact, many of these labels have a century-long history, so there is a certain fixed set of characteristics that parts with that label are *supposed* to have.

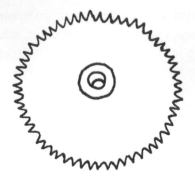

Figure 4.11 Meccano impostor gear wheel
Source: Lucy Charlton.

successful, but over time, with trial and error and lots of care and thought, their designs have got better and better. They learn not to use a part to try to do what seems not to have worked for themselves or others, and to duplicate uses of that part that have been successful.

Now if you build to their design you get, from Builder 1, a proper-looking motorcycle or steam locomotive; from Builder 2, a model of the Eiffel Tower and the Golden Gate Bridge; from Builder 3, a bulldozer and a windmill. Sometimes there's a hitch and the architect needs to adjust the design, maybe more than once, and every once in a while it just doesn't work at all. But in general, things go smoothly, especially when you don't ask for too much imagination. In this case we would happily order a model of something new, but not too new, like a Bugatti Chiron from Builder 1, a skyscraper from Builder 2, and a block-setting crane from Builder 3. And we would bet a reasonable amount they would come out well, maybe even with little or no fine-tuning.

Trust in the deliverances from the builders is not based just on their own past success but depends on much else. To begin with, they work from a shared pool of parts that others use too, and, though they do not have exactly the same conception of the characteristics these parts have, and, as postulated, they do not have a true conception, there is a strongly overlapping shared core. Second, they exchange ideas amongst themselves when they note something striking, though their account of what has happened may not actually be accurate. Third, it matters that there are a very great many parts—you can't build something interesting that is really secure without a good many very different parts. For instance, you can't build an orrery that will accurately simulate the solar system if you have only an elementary beginner's kit. Sometimes, even, you will need a particular special piece that may not be easily obtainable. For instance, with the extensive collection that a serious Meccanoman will have acquired over many years, you can build the classic 'Supermodel No.16: The Meccano Loom'—but not if your collection lacks the vital and hard-to-come-by 'Part 104: Shuttle for loom' (Figure 4.12).

The point is that all this can be done even though the parts may not be as the builders conceive them. Nevertheless, these parts together make up a good collection that the builders can learn how to use and learn what they can be used for—though the builders may not recognise what aspects they actually use. For instance, there are a number of part 5 pieces—the so-called '5 hole strip'—in Builder 3's windmill that stands erect and whose vanes turn round with the tiniest of pushes. Recall that the part labelled '5: perforated strip 5 hole' really has only three holes in the middle plus an indentation on both ends, not five holes as the builders believe. But without conscious recognition, Builder 3 only uses these pieces to put screws through, and for holding screws, the indentations on each end work as well as a hole. We might also notice, from our God's-eye point of view, that where five holes are needed other than to put screws through, Builder 1 never chooses part 5 but always uses a seven-hole strip (Figure 4.13). Recall also that '27a: gear wheel 57 teeth' is supposed to have a number of holes within its toothed circumference but in fact has no holes at all. But the lack of holes that are supposed to be there within the circumference of 27a would not have affected our friend Adrian's constructions when he was a young builder because, he reports, he never found a use for them in the first place.

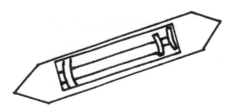

Figure 4.12 Meccano shuttle for loom
Source: Lucy Charlton.

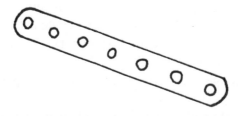

Figure 4.13 Meccano 7 hole strip
Source: Lucy Charlton.

This, we claim, is what science is like. There are lots of components necessary if science is to construct a device, model, measure, or any other scientific product that reliably does what it is expected to. The bulk of these components will be used in multiple contexts, where they may be differently understood but still the understandings share a family resemblance. In a variety of scientific domains including both the natural and human sciences, scientists have assembled such an assortment of components and learned how to use them in different combinations in different contexts to do different things, and they are able to do this even though the components may not really be what they are thought to be in any of these contexts.

In general, the supporting tangle for a successful scientific production should contain lots of pieces that also figure in support of tangles for scientific productions in other domains and that entangle with each other and the scientific production they help support in the 'right' way for the job. As with the Bugatti Chiron, the skyscraper, and the block-setting crane, when all these features are present, you have a far better chance that the new product will be reliable. And, returning to our comment from Section 1.4 that every labourer is worthy of their hire, we want to underline that all the different parts are necessary though you may get carried away in praise of one special one if it turns out to be just the one missing to do a special job you want—like the shuttle for the loom.

4.4 The 'Virtuous' Tangle: Three General Features

We philosophers all know that science involves a great amount of detailed hard work. But when it comes to evaluation, especially for scientific creations that are taken to be truth apt, like theories, we tend to picture the bulk of this work as consisting of the creation of more and more evidence. Perhaps we allow that something might be added about the motivation for a new hypothesis or theory—especially the problem setting which spawned it. But there is far more needed to support reliability judgements than what is usually thought of under the heading 'accumulating evidence'. It's the far more that we mean to catch in talking of the 'virtuous' tangle—a tangle of 'the right kind'—that supports reliability.

We propose three general features that a virtuous tangle will have. It will be

(i) *Rich*: this is meant to include not only that it has a lot of closely connected pieces but that these pieces are varied in type.

(ii) *Entangled*: the pieces relate to each other and to the product/aim pair in question in a variety of different ways.

(iii) *Long-tailed*: these pieces figure in support tangles for other scientific products in other domains that are succeeding at other kinds of difficult jobs, including successful interactions with the empirical world.

Obviously, these are neither necessary nor sufficient for reliability, and they are also abstract and imprecise (e.g. how many is many and on what scale?). But they are symptoms and you should beware if any of them is clearly absent.

(i) *Rich*. This means that the elements of the tangle of support for the reliability of a scientific production should be both many and varied. The expectation that many are required is at the heart of Karl Popper's ([1935] 1963) complaints about Freudian and Marxist theories and his insistence that proper science be falsifiable. Proponents of these theories might invoke their theory in defence of successful predictions and retrodictions about the world, but it would be they, not the theory, that made these predictions and retrodictions. These theories are not rich enough to do so, Popper complains. There aren't enough pieces in them. In particular, they lack the bridge principles necessary to link the abstract theoretical language with more concrete descriptions of behaviour. That is why they are not falsifiable.

Popper wants proper science to make claims that are so determinate in their meaning that they are falsifiable—no faffing about. What is being claimed is *exactly this*. If scientific claims, assumptions, and practices are highly determinate then there is a fact of the matter about whether you are using the same ones that have worked before. Without this, past success is little guide to future reliability.[11]

This has been echoed more recently by Christopher Hitchcock and Elliott Sober (2004) in their worries about theories that merely accommodate but do not really imply the data: if a theory is 'sufficiently plastic that it can accommodate any data that may come along, it is in no position to make predictions about what data will come along' (7).

Richness marks out not only the likely need for many elements, but the need for the right kind of variation as well. You see both these aspects of richness at work in the long list of different kinds of ingredients in the democratic peace tangle we discuss in Chapter 6: embedding in theory, concept development, construct validity studies, devising new measures, bridge

---

[11] Though Popper himself, taken up with David Hume's problem of induction, argued that even with this there is no rational basis for supposing the past will be a reliable guide.

principles, and so on. Although there are a number of well-developed competing theories of the democratic peace, each separately has all of these components. (And, looking forward to our discussion of entanglement, the components touch on each other and on features in other tangles in just the right way. For example, the construct validity studies support the concepts employed and the measures are provably good for those concepts.) This helps lock in what they are actually saying and/or doing: it is the same thing each time it is put to use to make predictions or interventions. Without this, previous successes with that claim or practice can provide little support for predictions of future success.

Take, as another example, the tangle of scientific products and knowledge claims employed in establishing the boiling point of water that we recount at greater length in Chapter 5. In the late eighteenth century, a committee of natural philosophers appointed by the Royal Society built upon a rich set of concepts, technologies, materials, practices, experiments, observations, and more to standardise the measurement of the temperature of boiling water. The pieces of this tangle of 'thermometry' included the concept of a fixed point (a natural phenomenon with a fixed temperature), the two-point system (the use of two fixed points rather than one in standardising temperature measurements), the Fahrenheit thermometer, the thermometric substance mercury (which the thermometer contained), the practice of dipping the thermometer into the water, conflicting observations of the consistency and inconsistency of the boiling point made in specific experiments, and so on. The richness of this background tangle allowed the Royal Society's appointed committee to state exactly how the temperature of a specific phenomenon was to be measured.

Crasnow's (2021) work on coherence objectivity shows a similar concern with what we call 'richness'. As Crasnow notes, the goodness of measures in the social sciences is generally assessed through 'reliability' and 'validity' tests (in a different sense of 'reliability' than the reliability this book is about). In literature on assessing measures, 'reliability' is used to refer to the consistency of a measure. A number of different categories of reliability are singled out, and although there is no entirely agreed-on schematisation, typically included are inter-rater consistency (which you have seen in Section 1.5 that PCRA and V-Dem are much concerned with), test–retest consistency, and the internal consistency of responses to different items on multi-item measures.

Validity is supposed to be the extent to which a measure measures what it is supposed to. That of course cannot be assessed directly. 'As a consequence', Crasnow (2021) notes, 'evaluation of measurement would depend almost entirely on...reliability' (1225). With respect to V-Dem in particular, she notes, 'Scholars affiliated with V-Dem have conducted reliability tests to determine to what extent various properties (for example, gender) of those experts might affect their reliability' (1221). Nevertheless, she argues, 'This kind of agreement among measurers does not in itself support the validity of the measures' (1221). For Crasnow, a great deal more is needed in addition. Recall her claim: 'It is a coherence of measures with theory, empirical knowledge, and practical knowledge that provides support for assessing the validity of the measures' (1224). Recasting this in the general frame we offer here: a tangle containing just reliability tests is not nearly rich enough.

Richness is also called for in our objections in Chapter 2 to restricting the reasons admitted in assessing policy effectiveness predictions to rigorously established evidence claims. A virtuous tangle to support the reliability of these predictions should include far more items and far more varied items. In particular, we argue there, it should include a causal-process-tracing theory of change and the large assortment of information called for in that theory plus the studies and investigations required to produce and assess those.

(ii) *Entangled.* The items that support the reliability of a scientific product shouldn't be like spokes on a bicycle wheel, centring each on the product/aim pair to be supported. They should relate not just to the product but to each other, and in the 'right' ways. We suppose that this is part of what Hasok Chang (2017) demands in looking for 'coherence' or 'harmony' among practices—the products they produce entangle in the 'right' ways to support reliability. Otherwise they may be constraining the pair in different directions at once and not narrowing down its features to get a fit, so that the identified product is reliable for the identified aim. What are the right elements to be entangled and what kinds of entangling there should be is a case-by-case matter. Still, for specific kinds of product/aim pairs there are specific kinds of elements related in specific kinds of ways that should generally be there.

We have already presented a mini example of this. Recall that the three most prominent indices of democracy (V-Dem, Polity2, and Freedom House) display a high level of agreement for those observations included in all three datasets, which is an almost essential ingredient in any virtuous tangle to support a measure's reliability. This kind of agreement plays a dual role. First, the requirement for agreement provides a direct constraint on what V-Dem is like. Second, as part of an extended tangle, it provides the basis for the constraint supplied by the kinds of empirical and theoretical studies that Crasnow discusses.

But Crasnow insists that various kinds of consistencies within a measure and with the results of other measures provide insufficient assurance that a measure is measuring what it is supposed to. Although she doesn't rehearse the reason, this is because, in the case of democracy, for instance, all the measures could be measuring the same feature, but that feature is not democracy as characterised. Or they could each be measuring a different feature, but these features are correlated with each other. Democracy as characterised may or may not be among these features actually measured. That is why the measure should have a great deal of theoretical and empirical support in addition. But the converse is also the case: the cross-measure agreement helps assure that the empirical and theoretical constraints are genuinely satisfied.

To get clear what we mean by this, consider a caricature example involving a new measure of democracy, new-Dem. Suppose a well-done study correlates democracy, measured by old-Dem, with economic growth—G (and suppose, to keep matters simple, that, contrary to fact, the measurement of growth is unproblematic). A second well-done study correlates positive results on new-Dem with growth. Both these can hold without there being any correlation between new-Dem and old-Dem.[12] But without a correlation between the two measures the two empirical studies have little to do with each other. For instance, although new-Dem is correlated with G, it may well not be correlated with 'old-Dem & G', so states that are both judged to be democracies by old-Dem and exhibit growth may not be any more likely than not to be classed as democracies by new-Dem. Each study shows that something or other is correlated with growth. A correlation between old-Dem and new-Dem provides reason to support that it is the *same* feature. Of course, to back up that that feature is *democracy* as characterised will take a much more extended tangle of support.

We can look to Chapter 2 for a second example of the kinds of relations entanglement might consist in. Recall that there we object to those in the evidence-based policy movement who insist on admitting only 'rigorously established' evidence for causal claims in the policy arena. These, we argue, are like the stiff little sticks in a Jacana bird's nest. They can't support anything much at all until they are woven into a rich tangle with lots of other different kinds of ingredients.

The simple lesson to take away from the Chapter 2 discussion is, as we note there, on richness. We urge in particular that to expect reliability from a claim about policy effectiveness in a new setting, RCTs elsewhere add little by themselves. An almost universally essential piece of work to support the reliability of a context-local policy prediction is the production of a causal-process-tracing theory of change for the policy, geared to that context, with

---

[12] For instance, here is just one among innumerable ways that this can happen: letting N = new-Dem and O = old-Dem, if P(NOG) = P(NO–G) = P(–NOG) = P(–NO–G) = 0.2; P(N–OG) = P(N–O–G) = 0.05; P(–N–OG) = 0; P(–N–O–G) = 0.1.

all the additional work that identifies and evaluates each of the numerous and varied ingredients that go into the process theory of change, some of which is at the level of general principles but much of which depends on the production and assessment of local knowledge.

It is not just important that these ingredients are numerous and varied, as we mentioned in discussing richness. Clearly they must also interrelate, and in the right ways. Consider the various items of theory that enter. First, the pToC for the programme should be a concrete instance of the abstract overall theory. Second, the causal principles for the different steps must be mutually compatible in that context: it must be possible for all of them to operate in that context and at the right times and places. So the work that supports those principles must be appropriate to the local circumstances. (Recall our discussion of the Vajont dam disaster in Section 3.4.2 where this kind of work was absent for the general principle that limestone reveals its flaws on its surface.[13]) Then the proposed support factors, derailers, and safeguards must relate to each other in the right way, and all of them must fit with how the causal principles are to operate in context at each stage.

It is not uncommon for ToC advocates to call for a list of 'assumptions' that must be satisfied if the programme is to produce the targeted outcome. But Cartwright et al. (2020) argue that lists are not enough. Lists do not make clear the different roles that different assumptions play and how they fit together. Flow charts like Figure 2.1 in Chapter 2 look complicated, which can put users off. But they do the important job of showing how the various ingredients that back up the policy prediction must relate to each other if a claim to reliability is to be warranted.

A third example is provided by the kinds of accounts typically offered in history and archaeology and sometimes in sociology, where the *narrative* is central in mapping out what pieces are required for support that an account is likely to be accurate and in constraining how these pieces fit together. Edna Ullmann-Margalit's (2006) discussion of prominent accounts of the Dead Sea Scrolls, *Out of the cave*, provides a good illustration.[14] By far and away, the most dominant one is the Qumran-Essene Hypothesis, which asserts that the scrolls belonged to the Essenes, a sect whose centre was at the nearby site of Qumran. During the last three decades, though, competing theories have emerged that deny the link between the Qumran establishment and the scrolls.

---

[13] Again, how much of this work needs to be in place for you to be warranted to act on a prediction depends on the different costs of different kinds of failures.
[14] The discussion here is taken from Cartwright (2014).

In *Out of the Cave* you see that different hypotheses about the Dead Sea Scrolls and the nearby settlement at Qumran are supported by different complex narratives. Are any of these narratives reliable for concluding correctly what the Scrolls are and how they got there? All purport to be backed by a good body of evidence. The supposition by advocates of each narrative is that the evidence available is enough for you to expect its conclusions to be correct.

Two important features of these narratives stand out. First, is how compelling they are; should they be true, they almost force their conclusions—as in a good argument. Second, from the point of view of one narrative, many of the facts adduced as evidence in the others are irrelevant—they only make a difference to the case *given* the story you are already in the midst of. If the Qumran site was the setting of a military fortress and not of a religious sect, then the fact that it is two kilometres from the place where the scrolls were located has no bearing at all on whether the scrolls are Essene or not. This illustrates our repeated claim that there is no such thing as evidence—there are just facts—until a vast body of scientific work turns these into evidence.

Particularly visible here is the importance of the entanglement: of how pieces of evidence relate to each other and how those already admitted as supporting the reliability of the narrative constrain what further can and cannot be added. *Out of the Cave* is a tale of thick narratives each involving interwoven strands of evidence, where each strand supports the hypothesis only because the others do so as well in connection with each other; and where without the rest of the pieces of the narrative, few single pieces have grounds for bearing on the reliability of the overall account. They are just dangling odd facts that may or may not be of interest on their own to someone or for some purpose, some purpose other than speaking to the adequacy of an account of the Dead Sea Scrolls.

(iii) *Long-tailed*. We begin by continuing our discussion of the Dead Sea Scrolls. Ullman-Margalit argues that the interplay of texts, including the Scrolls and classical sources (Josephus, Philo, Pliny), with archaeological findings form a 'strong linkage' argument in favour of the Qumran-Essene hypothesis. *Out of the Cave* makes a strong case that the support that backs the accuracy of the account is both rich (in her words, 'thick') and entangled ('interwoven'). Is it long-tailed? There's the problem. The tangle is closed in on itself, with few external connections and for many this undermines the case for reliability. Alison Wylie, for instance, adopting our idea of the tangle, describes it thus in an email discussion with us about it:

> ...this an extreme case – one in which a perfect storm of factors contribute to perversely circular reasoning: conflicting contemporary commitments overdetermine

what data will be gathered, how it will be interpreted as evidence and assembled, what background assumptions will be taken for granted and which will be scrutinized...in short, a vicious tangle.[15]

Robert Chapman and Wylie (2016), in their book *Evidential Reasoning in Archaeology*, argue that 'the scaffolding required to "see" material traces as archaeological data and to constitute these data as evidence' (84) include:

(a) High-level theory, for example, abstract framework assumptions about the nature of the cultural subject.
(b) Domain-specific models of particular cultures.
(c) Domain-specific models of the archaeological comparanda relevant to their investigation.
(d) Background knowledge bearing on specific elements of the archaeological record.
(e) The conditions that produced them that derive from an enormous array of fields, natural and cultural.
(f) The technologies and institutional infrastructures by which orienting 'theory' and background knowledge is internalised, leavened by experience and translated into practice.
(g) The trained skill and judgement that helps with this. (85)

Among these, (a), (d), and (e) are especially likely to introduce the variety of long tails that can thwart perverse circularity and help build virtuous, not vicious, tangles.

This support for the Qumran-Essene hypothesis contrasts with the case for V-Dem that Crasnow (2021) rehearses. Each of the five constituents of the *Ballung* notion of democracy that V-Dem includes have long tails stretching widely into philosophy, political science, history, religion, and economics. Witness, for instance, the large number of academic journals with some variation on the word 'democracy' in their title. Apparently, many Chinese citizens would like to revise these, Crasnow notes:

Surveys indicate many Chinese perceive their form of government as democratic and, even more surprising, do not take elections to be a key element of democracy. The understanding of democracy underlying these views is that a regime is democratic if it governs in a way that is consistent with the well-being of the people. (1220)

---

[15] Email to Nancy Cartwright, 20 October 2020.

But, she argues,

> ...it is hard to see how China would count as a democracy by any measure developed through the use of the dataset and consistent with their five principles. In fact, China ranks near the bottom on all five indices of democracy (2019 V-Dem Annual Democracy Report). It ranks near the bottom on Polity IV and Liberty House indices as well.   (1221)

We have already seen though that agreement among indices goes only a small way to support the reliability of any of them to indicate what they are supposed to. Crasnow, as we have seen, locates the solution in the extensive body of empirical and theoretical work on democracy done outside the arena of measurement: '[I]t is possible to argue that China does not count as a democracy for theoretical reasons—the Chinese concept of democracy is not compatible with the body of democratic theory' (1227). It is the long tails that support the reliability of a measure of democracy that excludes governing in a way that is consistent with the well-being of the people as one of democracy's central features.

Alongside these case studies, Rens Bod et al. (2019) have introduced a concept that resembles, though is not identical to, that of long tails. To better account for the interaction between disciplines, the authors propose that 'cognitive goods' can 'flow' across fields of study, allowing for interdisciplinarity. Like ourselves, Bod et al. are concerned with the various products of science. Cognitive goods can be 'virtues, institutional practices, concepts, formalisms, methods, material objects, and metaphors' (487). Yet, in contrast to long tails, which are established connections between different domains of knowledge, cognitive goods are what initially circulate and facilitate the creation of further reliable products. Perhaps the exchange of cognitive goods, then, is one mechanism by which long tails form.

In Chapter 5 our central illustration of the kinds of roles long tails play is in the development of penicillin. But there you also see the importance of long tails in the establishment of the boiling point of water. An important part of this story is that the tangle of thermometry continuously developed long tails into a tangle of barometry. For instance, Daniel Gabriel Fahrenheit found that the temperature of the boiling point varies with atmospheric pressure, and the Royal Society committee appointed to determine how to standardise measurements of the boiling point produced similar findings (Cavendish et al. 1777). This led them to the first rule of their preferred method: that the boiling point should be measured under a standard pressure of 29.8 English inches (~757mm). Stretching even further, both the tangle of thermometry and the tangle of barometry were embedded in the larger tangle of meteorology since thermometers and barometers were the instruments of meteorologists (Cavendish 1776).

Before moving on, we should note that, just as with richness and entanglement, we don't expect to be able to find general criteria for what makes for sufficiently long, sufficiently good tails: How much is enough, what are the right kinds, into which aspects of which other domains? But this does not mean that there aren't good reasons case by case to judge some as deficient and others likely to be adequate.[16]

4.5 What's in a Tangle?

What, as we conceive it, is in a tangle? Section 4.3 pictures the constraints imposed by a tangle to be a central mechanism by which tangles underwrite reliability. Here we fill in more detail on just what the constraints imposed by a virtuous tangle could do that makes for reliability. We then use that to determine what needs to be in a tangle. Our task is virtually complete once we have done the first since what we have to say at the second stage is almost a truism: what needs to be in a virtuous tangle is whatever it takes to impose the requisite constraints. We approach the question from two different perspectives, offering two different answers, one from the point of view of nature's practices, the other from the point of view of the practices of the sciences.

Begin from nature's point of view. Why is scientific product P reliable with respect to aim A: why is it reliably able to do A, which we aim for with it?[17] This is a question about the way the world is. It has a purely empirical answer. This depends on the facts—the facts about P and about A and about the world and about how the P/A pair relate to the world. When P is truth apt, for instance a theory, a claim, or a model, this is the kind of thing that the scientific realism debate in philosophy of science is about. Suppose P is a model and A is a prediction about an outcome O that we wish to be true.

---

[16] Can virtue come in degrees? Surely. But we expect you can get at best a partial order. If you start with a given tangle and make it richer, or more properly entangled, or more long tailed with respect to a given purpose, it will become more virtuous with respect to that purpose. But in general there is no way to trade off or compare improvements along one dimension versus along another. And we see no way to compare different tangles where dominance is not available to use, though sometimes it will be clear that one tangle is so good along all three dimensions and another so bad that the first is clearly more virtuous than the second.

[17] It is worth a quick reminder that A encompasses both internal and external aims for the project. It is not enough to simply say that you wish to measure X; one must also specify what purpose you are measuring it for. The same goes for any other product of science.

One satisfactory answer is that P pictures causal processes by which O is produced and does so correctly.

So, by 'What is in a tangle "from nature's point of view"?', we mean what is in a tangle that is genuinely conducive to reliability. If a scientific product really is reliable, that is, it works repeatedly and regularly in the way you use it to do what you aim for, there must be a reason for that, a reason based on what obtains in nature, though the reason may not be anything like what you take it to be. That's why we use the expression 'from nature's point of view'.

We are thus supposing that there is a fact of the matter about which things promote others in the empirical world. If some facts hold, then others are likely, where likely is not a formal probability notion but reflects facts about nature itself, its patterns and laws. We suppose that nature is not entirely capricious: there are consistent, intelligible, and systematic patterns even though science may never recognise correctly what they are. This seems required if successful prediction based on the accumulation of reliable scientific products and knowledge claims is to be possible, as we know it is. It is not just random chance that lasers built according to a particular model emit the beams predicted. As we suppose in Section 4.3, they do so because of the features the parts actually have when assembled together and the principles that actually govern them in that arrangement, even if these are not as science supposes them to be.

So our answer *from nature's point of view* is almost a truism:

What must be in a tangle that genuinely supports the reliability of a scientific product is enough constraints on the make-up of the product and on the way it is used on the one hand and on the aims it is used for on the other so that when you use that product that way you are doing something that promotes that aim under the principles that nature operates by given the empirical facts that obtain.

The idea here is familiar from philosophical discussions of supervenience. S supervenes on S′ if S′ ensures that S obtains but not necessarily the reverse. Some definitions make 'not the reverse' a necessary condition.[18] When science performs what it takes to be P (plus perhaps something it does not recognise it is doing, X) it must be doing *something* involving features that really obtain in nature even if science's conception of what it is doing is wrong. And if P + X regularly and repeatedly yields A, then that something that science is doing must evolve under the laws of nature into something that science

---

[18] For more on supervenience, see Bennett and McLaughlin ([2005] 2018).

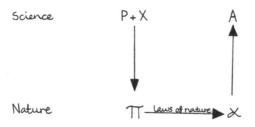

Figure 4.14 Securing reliability from nature's point of view
Source: Lucy Charlton.

recognises as A, even if the description of the evolved state as A is also wide
of the mark (Figure 4.14). So when product P used in the way it is used reliably serves aim A—where P and A are identified in the conceptual scheme
used in the relevant scientific community—there must be a corresponding
π and α and maybe an X—where π and α are in terms of the true features in
nature—such that

π supervenes on P + X[19]
α is produced under the laws and principles of nature given π[20]
$\alpha \to A$

Why X? No matter how finely science constrains P, there will always be other
things true as well when P obtains. Sometimes there are things that go along
with doing P most all the time, and these may be necessary to ensure π
obtains. X refers to these. You can't do P + X without π but you can do P
without doing π. X needn't be something that science must do to do P nor
that science always does; what matters is that A is served when X is there and
not reliably otherwise. Noticing failures and trying to fill in a missing X is
standard procedure in science.

But success in identifying an X that works needn't imply truth. Consider,
for example, past methods from psychology. In the second half of the twentieth century, experimentalists stimulated different parts of animals' brains by
implanting electrodes and then delivering a current (see e.g. Olds and Milner
1954; Olds 1969). Although these methods produced reliable behavioural
effects—say, positive reinforcement as measured by an increase in lever
presses—neuroscientists today maintain that spatially localisable areas

---

[19] P + X could on different occasions correspond to different states of nature that all evolve into states
of nature that science recognises as A. We can allow that possibility though it is many times farfetched.
[20] We put it this way rather than '$\alpha \to \pi$' to avoid a commitment to determinism. If α leads to π by law
but not universally, P + X will both 'regularly and repeatedly' (i.e. reliably) yield A though it won't do so
'always'.

actually contain spatially intermingled but functionally distinct populations of neurons (Morales and Margolis 2017; Li 2019). Hence, a function previously localised to a given brain region via reliable methods of electrical stimulation is merely the net output of activating a heterogeneous collection of cells, the subsets of which differentially subserve different functions. Indeed, contemporary neuroscientists now have reliable methods for producing different and sometimes opposite behaviours by targeting different sets of genetically defined cells (Cardin et al. 2010; Faget et al. 2018; Wilson et al. 2019).

For science studies scholars, our emphasis on nature's point of view may look familiar. Proponents of actor–network theory have long drawn our attention to the 'agency' of non-humans. According to these theorists, scientists are not the only ones responsible for what happens in science. By 'making some difference to the state of affairs', as Bruno Latour (2005) puts it, objects are actors too (52). Along these same lines, Andrew Pickering (1995) has focused on the role that machines play in how scientific practice unfolds in real time, arguing that there is a 'dance of agency' between scientists and their instruments (21). Scientists work towards achieving their aims, but the achievement of these aims depends on what the materials they use do as well.

This answers our question about what must be in a tangle if it is to support reliability. But science cannot see directly into nature to see what from nature's point of view is needed, let alone what to include in its tangle of work to ensure reliability. So what does science try to put in a tangle in order to support its expectation that a product will be reliable?

To answer that it is worth revisiting reliability. We have stressed that reliability is a two-place relation: P in the way it is used is reliable for A. And we have underlined that a major job of science is to ensure that the products it creates are reliable to do what is expected of them. Science is in the business of creating new things all the time, new products—measures, models, theories, concepts, narratives, studies, devices. It does not create these just for the fun of it, like architectural follies. Rather, these products are supposed to do something. So creating a new product is a Janus-faced enterprise: there's the product on one side and what's to be done with it on the other. Both need delineation. To assess the reliability of a product, then, you need to know just what that product is and how it is to be used, and you need to know what it is expected to do. Neither of these is ever genuinely explicit—it couldn't be if you take seriously meaning as use in the context of science, as we urge you should. It involves, instead, a great deal of implicit understanding (as we stress with respect to purposes in Chapter 3).

Since so much of our discussion otherwise focuses on products, here we will look specifically at aims. Meaning as use aside, it is hopeless to try to state these explicitly without the use of major fudge words. This is equally true of zeroing in on what the product is, but in the case of aims this lesson is even entrenched in folklore. This is one of the take-home messages of the old tale of the monkey's paw,[21] where an old couple are granted three wishes. The first two are granted as requested, but with unintended add-ons. The couple get the money wished for to pay the mortgage but it is in recompense for their son's sudden death in a factory accident. The final wish—to get him back alive—is then granted when he returns to live in his mutilated and decomposing body.

The couple could have been more explicit. But it is clear that no matter how much they added, the wicked fakir who cursed the paw can always grant what is asked for but is definitely not what is wanted, if only the fakir is clever enough. Unless, of course, the couple add something like 'and it is exactly something we really would want', which defeats the enterprise of being explicit. Nevertheless, we all know that what the fakir provided was definitely not what the couple wished for, and also there are a great many versions of 'receiving money for the mortgage' and 'getting our son back alive' that we would all accept as within what they had wished for. The couple, the fakir, you readers, and we authors have a shared understanding of what is and is not aimed for, albeit with fuzzy, possibly very fuzzy, edges.

We spend considerable time in Chapter 3 talking about aims vis-à-vis objectivity. It can be helpful to recall that discussion here. We argue that objectivity in the loose *Ballung* sense is often employed to enjoin OTBF, and that often what is to be found in the case of a scientific activity, like designing the Vajont dam, includes the appropriate aims for it. The full set of appropriate aims is not explicitly given, in analogy with the full set of what you are required to do under a duty of care. The injunction can be fair nonetheless because there can be sufficient constraints on what are and what are not appropriate aims—some things are definitely in, some out, others negotiable. In Chapter 3 we use the term *context* to refer to the body of features that fixes these—the facts of the case at hand, including empirical facts, norms, habits, and customs, currently available knowledge (both knowledge that and knowledge how), and a shared understanding about expectations.

---

[21] Here we refer to 'The Monkey's Paw' by W.W. Jacobs, which was first published in England in the collection *The Lady of the Barge* in 1902.

Analogously, when science creates a new product that is to be put to use, it must ensure that there are sufficient constraints that, within bounds, fixes what aims are appropriate for it in the relevant contexts. Where there are insufficient constraints to do that, pieces must be filled in until a shared understanding is possible.[22] So, the tangle that supports the reliability of a product of science must include enough constraints to provide an understanding of what are and are not aims that it is supposed to be reliable for.

The process of creating a scientific product, then, involves the creation of a paired understanding of what the product is and how it is used, on the one hand, and of what it can reliably be used for, on the other. What we are calling the tangle of products that science supplies to support reliability provides the source of this understanding. These constrain, on the one hand, just what the product is like and how it is to be used and, on the other, what it is to be used for.

Science must stabilise P and the ways it is used, and it must stabilise what A is as well, and it must hone them to match. You would seldom want to bet on a product to produce A reliably if you had no evidence that it had already done so a number of times. Stabilising just what P and A are ensures that is the same P and the same A in new cases as in the cases where P has worked before. Or at least it ensures that they are the same from science's point of view. Nor would you want to bet on P's reliability for A failing good scientific reason why it should be able to result in A, where that reason might well include among much else a record of past success. This leads us to the conclusion that

From science's point of view, a virtuous tangle should include elements that constraint the nature of the product and the way it is used and its aims.[23] These constraints should be tight enough so that when the product satisfies these constraints it can, given the available knowledge, reasonably be expected to be reliable for satisfying what the constraints delineate as its aims.

Here we are thinking of the elements of the tangle performing a job similar to the one Andrea Woody's (2015) 'functional account' assigns to explanatory practices:

---

[22] Who must share the understanding? That depends on the problem to be solved and on the users. This can create serious dilemmas when it comes to practice. For instance, social scientists are often taught statistical techniques that they misuse because they do not share the requisite understanding of the aims to which these techniques can be put. What is the appropriate remedy for this? We invite further thought to see what advice can be offered.

[23] Recall that there can be very different aims expected to be served, either all at once or different aims in different settings.

[The] production of explanations serves to *constitute* . . . "intelligibility" for a scientific discipline. . . . [I]t functions to sculpt and subsequently perpetuate communal norms of intelligibility.   (81, original emphasis)

For Woody, explanations constitute norms of intelligibility in science. Analogously, in our case the elements of the tangle constitute what a product and the aims it is expected to satisfy are—what things are instances and what are not. Consider the PCRA discussed in Section 1.5, which is supposed to predict recidivism. Part of what constitutes this as a product for predicting recidivism are studies in which the PCRA seems to have successfully done so already in a variety of places (though of course given our discussion of variation in Section 4.1 there had better also be good arguments available that the variation is good enough to go on). But as we note in Section 1.5, the scientific community would not judge the PCRA a product appropriate to achieve this aim (or the aim an appropriate one for this product) without a good study on inter-rater consistency. Without that there is little assurance that it is the PCRA as it is intended to be used that produced the good predictions in the past cases (if it really did so) rather than particularly insightful raters. It is only the 'PCRA algorithm as constrained by past studies of success + . . . + a good inter-rater-validity study' whose reliability is up for evaluation.

We can think of the honing of the characterisation of the product and the set of aims it is expected to serve to get a matched pair, <P,A>—a product used as it will be that promotes those aims—in the formal mode in terms of an argument. A virtuous tangle contains the elements that support a good argument that P used in that way will regularly and repeatedly yield the matching A. What is and is not a good argument, though, depends on what standards of relevance are adopted. These, we take it, are set by the scientific community. Here again we parallel Andrea Woody's (2015) arguments about standards of explanatory relevance:

Relevance is stipulated based on communal norms. As a consequence, individual scientists rarely make independent decisions regarding explanatory relevance; with proper training, they can make reliable judgments regarding what is relevant and what is not, but this is primarily a matter of recognition, after social conditioning, rather than independent determination . . .   (81)

We know, though, that different scientific communities have different standards, which is something that Woody also stresses:

> The functional perspective... provides some resources for understanding the distinct dynamics of inter-disciplinary versus intra-disciplinary disputes regarding whether a given phenomenon needs explanation and if so, what would be required...
>
> ...[T]his account assumes that within communities, there are significant commonalities among accepted explanations. Heterogeneity may appear across practices but should not be expected to occur routinely within them. Furthermore, embedding and bridging relations between scientific communities and sub-communities will place substantial constraints on the explanatory diversity that can be accommodated across them. The fact that communities must interact and work together entails that their models of intelligibility must be generally commensurable.   (83)

The models of intelligibility of different scientific communities might be generally commensurable as Woody claims, but, as we all know, that does not mean that the judgements from within each will agree on what is and is not a good argument for reliability. What one set of standards may judge to be a well-supported <P,A> match, another may judge to be poorly supported, or worse, judge that P is in fact a poor way to achieve A—what one takes to be a virtuous tangle, another might take to be vicious.

Drawing upon the work of the historian A.C. Crombie (1994), the philosopher Ian Hacking (1994) has made a similar point about so-called styles of reasoning being 'self-authenticating'. Since a particular style of reasoning sets the standards for what counts as evidence, it is only according to that style that we can say a statement is true or false. This kind of conflict makes life difficult when you want scientific advice about how to get A. But it does very often force the advocates on different sides to look for holes in the arguments of the opposition and to tighten their own, as we see in studies of the democratic peace and of gravitational waves discussed in Chapter 6 and in our Afterword.[24]

We have been talking throughout about *virtuous tangles*. With the distinction in hand between nature's point of view versus that of the sciences we can now clarify our use of the term 'virtuous', which was willed to us by Alison Wylie (recall her illustration of what she regards as a vicious tangle in Section 4.4). Our claim is that, above and beyond the usual suspects, virtuous tangles are conducive to reliability. We mean this to be something we

[24] Which is the relevant community then? Recall that we are answering the question, 'What is in the tangle from the point of view of science?'. The relevant community is the one intended in the question. There will be different answers depending on which is chosen for attention.

defend and illustrate by reference to investigations of what science is doing when it is successful and what can be identified as faults when things go wrong (as in Section 4.2). So we are talking about good—virtuous—tangles from a point of view within science.

There is, of course, a perfectly good sense of 'virtuous tangle from nature's point of view'—a tangle that in fact constrains science to do what will work in nature. But it is possible to have constraints that manage to mark out product/aim pairs that in fact work but that science does not see as such, and the converse: science supplies the ingredients for a really good argument that P used as it is used will yield A when in fact it will not, as we just noted. It is because of this latter that we claim at most that a virtuous tangle is conducive to reliability, not that it in any way ensures it, a point that is underlined in our discussion of gravitational waves in our Afterword.

4.6 What's So Good About a Virtuous Tangle?

We have, we believe, answered the twin questions of what is in a tangle from nature's point of view and what from science's point of view. But there is a lacuna, a big one. Our account describes what science must be doing when it is reliable and also what science does viewed from within, from its own perspective. But we do not show why it is that when science does what it should from its own perspective that so often results in it doing something that nature demands for success.

The most straightforward answer to what allows science to produce products that nature says yes to, would be that science generally gets it right about what nature is like. That answer won't do. We don't mean to insist that science usually fails to get it right but, as we stress in our arguments for focusing on reliability in Section 1.5 and again in Section 4.3, it is well known that a panoply of scientific products have been, and many still are, reliable for the aims demanded of them despite the fact that current science deems them wide of the mark, not just discarded theories and concepts from past successful science but hosts of idealised models that don't even look accurate to the facts, measures that give the right answers but for the wrong reasons, causal process studies that identify a number of required factors correctly but do so by supposing the wrong set of causal principles to be at work, and so forth. What's to be said about these?

Before turning to this question, it is worth explicitly separating it from a similar-sounding question that we won't discuss: why are you warranted in expecting P to do A? This is a question about warrant, entitlement.

Sometimes the answer is just 'someone I have every reason to trust told me.' Sometimes that is not enough. The question at hand is different: what is it about science that is responsible for its producing so many reliable products? Or, more specifically, what is it about science, and especially about the methods, tools, and practices used to produce P and delineate A that makes P reliable for A? Or, since we don't really think anything will secure reliability for sure, what is it about the methods, tools, and practices used to hone P and A that makes them *conducive* to P's reliability for A? This is our question.

This is where the usual suspects are offered in answer—scientific products are so often reliable because science is objective, rigorous, and works in accord with the scientific method (or in accord with a host of sound methods). We agree that all three, and many more, are often conducive to reliability. But, we argue, there is an elephant in the room that they do not mention—the virtuous tangle of other work that underpins these methods, practices, and tools, which both gives a settled sense to them and, we maintain, is independently conducive to reliability.

That again brings us to the question, why is a virtuous tangle conducive to reliability? We have a few things to say that seem relevant, but we should confess from the start that we cannot answer this question. Nor can anyone else answer it for any other account of what makes science as we do it so often eventually successful, at least not without big metaphysical assumptions and even then often the answer is too sweeping. That's because an answer really involves solving the problem of induction: why do your favoured methods for learning about the world work?[25]

For David Hume, the favoured method was inferring that the future would be like the past, possibly on the grounds that past futures have been like past pasts, which as Hume pointed out doesn't get you any forwarder unless you presuppose already that the future will be like the past. Beyond that, this is not much of a specification of a method—like in just what ways? Similarly with the many 'epistemic virtues' that good science is meant to have, like parsimony, which we discuss in some detail in Section 1.2. Why should parsimony be a guide to genuine knowledge? Because nature is simple? How do you know that? And, of course, there is the problem of just what kind of simplicity does nature exhibit if it is simple? Or another obvious requirement we'd like a virtuous tangle for a scientific product to meet: test, test, and retest that it does what it should, or really, that it has done what it should. Why does that mean it is likely to do the same again? Especially since the best you can do is require that the same product has satisfied the same aim by your

---

[25] As Robin Hendry pointed out to us.

standards of sameness, which you cannot assume are the same as nature's standards of sameness.

Once we consider more concrete methods, answers can become more concrete. Recall our discussion of RCTs in Chapter 1. There you see a clear answer to 'Why is an RCT reliable for unbiased estimation of an average treatment effect?'. But this answer is not without presupposition, both major metaphysical ones (e.g. cases under consideration are governed by principles that can be represented by potential outcome equations) and, as Elliott Sober stresses (recall Section 1.2), 'low-level' substance-specific ones (e.g. those that rule out all post-assignment confounding). Similarly, for specific tangles for specific <P,A> pairs we can make specific arguments that the tangle is conducive to reliability that are reasonably compelling. Equally, for specific cases we can raise specific, reasonably compelling reasons to be wary. But the arguments always depend on other assumptions, and if you do not suppose these other assumptions are right, the arguments are no longer compelling.

That said, the network of heavy constraints on both P and A that a virtuous tangle provides offers some reasons not available to other accounts for why those products of science so often work even if science does not have the correct understanding of why. Part of the standard requirements under almost any account of the successes of science, which we too expect in virtuous tangles, is test, test, retest—especially with 'severe' tests, that is, tests of the product in conditions where you most expect it to fail if defective as opposed, for example, to those where you most expect it to work. As in what in causal process tracing are called 'smoking gun' as opposed to 'hoop' tests (see e.g. Van Evera [1997] 2015).

So, we suppose that most virtuous tangles include studies that judge that what you call P, or something very like it, has produced something you are willing to call A, or something very like it. The tightness of the constraints on what P and A are, then, seem likely to ensure that when you produce what you call P again, it is 'the same' as what you did before and so should be able to produce A again. For one-off products, like the Vajont dam, the next best thing is demanded: make sure all the pieces in the supporting tangle are severely tested for their capacity to play the role required. This, too, goes a long way to ensuring that what the product is is securely pinned down, at least from science's point of view.

This reason is not so compelling as one might wish, however. For the constraints provided by all those interrelated elements of the tangle work to ensure that P and A are the same from science's point of view. As we have noted, that need not be from nature's point of view, and it is nature's point of view that matters to what actually transpires.

Still, the richer the net and the more its elements are entangled in the background tangles for other products that work for their designated aims, the fewer options there are about what facts of nature P and A pick out. What's needed for what you call P to be reliable for A—that is, for what you call P to be able to repeatedly and regularly produce what you call A—is that P usually picks out that same π that will produce α, or something near enough to π to produce something that science judges an A. Either it's P that picked out π where P worked, not the unrecognised X, or where X mattered (as we suppose may often be the case), the usual practices for using P turn out to ensure X, though unbeknownst to you. The fact that the other products themselves have been judged in empirical studies to work for their objectives in combination with still others means that they too, on these occasions, must have 'corresponded' to the facts in the way pictured in Figure 4.14. All this puts very strong constraints indeed on what the overall pattern of real facts looks like.

We think of this figuratively as 'nature getting herself locked in'. Recall the example of the model of the helium-neon laser from Section 4.3, where there are hundreds of detailed assumptions, concepts, other models, measures, and more that cooperate together to allow the model to predict that a 633nm beam will be emitted. Each of these appears in the tangle for products that, in the really good cases, have themselves been judged in empirical studies to have successfully fulfilled their aims, though for each of these elements and for each of the new products, the element cooperates with a new group of other highly detailed elements. And each of the elements in each of the new groupings, like each of the original elements, participates in myriad other successful products. And so on. All this constrains more and more what you do when you do P and judge A. This locks nature in more and more about the reasons that that very thing worked once but won't regularly: In what ways relevant from the point of view of the laws of nature does what you did when P worked differ, from nature's point of view, from what you do when you do P next time and the time after?

No guarantees of course. It is an everyday occurrence in science that things backed by what are reasonably judged virtuous tangles that seem to have worked once do not do so reliably. Which should not be surprising, since whatever science does it inevitably does it from science's—not nature's—point of view.

What do you do when a tangle is not so virtuous? When there are holes in a tangle? Or when products that fill the holes are flawed? Or when products are poorly labelled, with no explicit claims about just what they do or where they work and no good body of practice to fix this implicitly? Or when the specifications for a product are sketchy so it is not clear just what the product is? Or when different schools of science disagree about questions like this?

These are all common occurrences in science. After all, much of science is fuzzy and imprecise, full of conflicts, disagreement, and shifting, unarticulated, or disputed aims.

Of course, often what a product is reliable for is vague or difficult to articulate, or there are gaping holes in the tangle that supports that it can do what is on the label. This doesn't mean there are no aims it is meant to fulfil or that there is no tangle there at all. Even the most undirected scientific observations require some idea of what they are producing and what it should do, and of how and why they expect it to work. Without some delineation of what you are doing and what it is for, no matter how vague, it seems difficult to claim that what you're doing is science at all.

It might not be that extreme. In many cases the uncertainty creeps in not because scientists have no idea what they are doing or why, but because there are genuine questions about, for example, how to decide which of two conflicting experimental results to include in the tangle, or what the precise aims of this project are, or what constraints should be prioritised where one is forced to pick between several viable alternatives. Such cases are common throughout science—think back to the competing theories about high-temperature super-conductivity discussed in the Preface, where the two sides placed different constraints on their tangles based on differing interpretations of experimental data.

What do you do in situations like this? Well, what do you do when you are thinking about truth, or reliability, or acceptance, or whatever, not in terms of the tangle but in terms of the usual suspects? When methods of confirmation fall short of rigour, as RCT advocates suggest is the case with respect to all other methods for confirming causal conclusions (recall Section 2.4)? Or when instruments are accused of political bias, as in Worsdale and Wright's worries about objectivity (recall this discussion from the Preview to Chapter 3)? Or when there is dispute about whether a method is up to standard (e.g. the method of dipping a thermometer into water to record the boiling point, which you read about in the Chapter 5), or whether a method can deliver what you want, or is applicable in a particular case (as in Sober's discussion from Section 1.2 of the principle of the common cause and the gene that may or may not make individuals infertile after the first child)?

Dealing with worries about a tangle is just like dealing with these other worries—it is business as usual in science. We don't have anything especially new to add here. But we can remind you that there are a few standard practices to fall back on:

1. Narrow the domain of application. Where there's uncertainty about the reliability of a tangle for a particular purpose, because it's unclear how to assess the virtues of the tangle's components, because it's unclear exactly what the aim of the project is, or where scientists are genuinely unable to articulate their goals or support their work with a rough tangle to back it up, then there are grounds to think you aren't doing the project you think you are. One standard move to make in these kinds of situation is to back up a step and try to restrict your work to constrain it to what you do, in fact, know.

2. Do more science. Adding to uncertain tangles, whether by doing more experiments, revising theories, using new techniques, measures, data categorisations, or equipment, looking for useful connections to draw to other work, or finding more constraints, amongst other things, is one of the primary ways science progresses. The building of increasingly virtuous tangles is a process that happens over time, often with many revisions, errors, and backtracking. It forms, however, the basis of many scientific advancements.

3. Hedge your bets. Sometimes you're forced to make a choice about what scientific product to use and, because of time or other resource limitations, you cannot undertake further work to help you decide between the two, nor can you narrow the scope given your goals. In these contexts it seems sensible to hedge your bets or run some form of risk analysis. This move tends to involve a widening of considerations using factors from outside the prior scope. You might consider, for example, minimising damage to the environment or considering the cost to the taxpayer of deploying a technique to help decide between products with fuzzy or uncertain tangles.

It is one of the strengths of the tangle account that it can help us articulate why these strategies are common in science, as in life. Many of the examples in this book are fairly clear cases of virtuous or non-virtuous tangles because that is the easiest way to demonstrate our account. Non-ideal cases are far more common, however, as you will see in the next few chapters where we draw on examples from the natural, biomedical, and social sciences to demonstrate examples of tangles that conflict due to different prioritisations of evidence, explanations, and theory.

5

Illustrating the Tangle: Episodes from the History of Science

Preview

In the last chapter, we discuss why virtuous tangles are both symptomatic of and conducive to reliability in science. We now turn to two detailed illustrations of virtuous tangles from the history of science—one that is likely familiar to readers and another that may be lesser known. In the first historical episode, we examine how natural philosophers of the eighteenth century 'fixed' the boiling point of water. In the second, we shift to the twentieth century to analyse the development of the antibiotic penicillin. For the first, we make use of Hasok Chang's (2004) *Inventing Temperature*; for the second, we call on Robert Bud's (2007) *Penicillin*. We supplement our analysis of these narratives using primary sources.

Before moving forward we should briefly clarify how our retelling of these past events connects to earlier arguments in the book. By using the term *episode* instead of *case*, we follow Chang (2012) in trying to avoid an inductive use of history in philosophy. Although we have consciously selected which episodes to include, those that are included are not meant to offer particular instances from which we would like to generalise about the nature of science. That, after all, would be contrary to familiar cautions about generalising and induction. Rather, our argument is that there are times when scientific work is reliable for a particular aim and that virtuous tangles are conducive to such reliability. The following episodes are meant to give you a vivid sense of what our abstract ideas amount to in some real concrete episodes, to show that these ideas make sense when they confront the world, and to demonstrate that they have real bite.

In light of these aims, we propose that the specific episodes we have selected are *canonical* for our philosophical project. According to Agnes Bolinska and Joseph Martin (2020), a 'canonical case is one that can be explained by factors relevant to the philosophical question at hand—and in

The Tangle of Science: Reliability Beyond Method, Rigour, and Objectivity. Nancy Cartwright, Jeremy Hardie, Eleonora Montuschi, Matthew Soleiman, and Ann C. Thresher, Oxford University Press. © Nancy Cartwright, Jeremy Hardie, Eleonora Montuschi, Matthew Soleiman, and Ann C. Thresher 2022. DOI: 10.1093/oso/9780198866343.003.0006

which the outcome is sensitive to those explanatory factors' (43). Here, the three features of a virtuous tangle—its richness, entanglement, and long tails—help to explain both how the boiling point was established and how penicillin became a medicine. Yet, the canonicity of our episodes is necessarily tentative, as the contingencies of the past are always subject to debate. With more research, it could very well change. Indeed, for all historical episodes (or case studies), canonicity is a 'negotiated property', 'established through a give-and-take between historians and philosophers' (Bolinska and Martin 2020, 42). Thus, the relevance of the following episodes to our aims is unavoidably a function of the historiography and primary documents we rely upon.

Of course, Chang and Bud do not themselves enlist the tangle as an analytic tool. For Chang (2004), the establishment of the boiling point illustrates a kind of coherentism that he calls 'epistemic iteration': 'a process in which successive stages of knowledge, each building on the preceding one, are created in order to enhance the achievement of certain epistemic goals' (45). Temperature standards evolved in this way, he argues, with bodily sensations of hot and cold first being replaced by thermoscopes (instruments that measure temperature ordinally), then by thermometers, and later, by a standard method of recording the boiling point using steam rather than the water itself. Although our episode on the boiling point involves the iterative process Chang describes, we also see in his narrative something else, namely a virtuous tangle. We argue that it is a virtuous tangle and the constraints it imposed on natural philosophers—or rather, they imposed on themselves—that helps explain why they perceived their standardising method as reliable.

This argument also points to another mechanism underlying epistemic iteration, and hence, the progressive coherentism Chang endorses. The 'principle of respect' (for past standards) and the 'imperative of progress' (relative to one or more epistemic goals) make possible the 'enrichment' and 'self-correction' of a system of knowledge, even in the absence of a secure foundation. As we show below, epistemic iteration also depends on the tangle. Without it, there would not be past standards to respect or aims to make progress towards.

In Bud's reconstruction of the invention of penicillin as an antibiotic, he emphasises techniques developed by Alexander Fleming as well as the multidisciplinarity of Howard Florey's research group. We see the former as allowing for rich and entangled products within bacteriology whereas the latter was a precondition for the creation of long tails. As with the laser cataract surgery we discussed in the Preface, medical treatments such as penicillin show why connections between various domains of knowledge are necessary.

5.1 Boiling Point(s)

To begin, we present readers with an overview of why and how natural philosophers sought to establish the boiling point. We then turn to an analysis of this episode using the concept of a virtuous tangle.

In *Inventing Temperature*, Chang (2004) traces the history of thermometry from the seventeenth to the nineteenth century. As he explains, those engaged in various temperature-measuring practices in the late seventeenth century faced a common problem. Although thermometers were used as early as 1600, they were not yet standardised.[1] Thermometers could vary in the temperature-sensitive substances they contained (air, mercury, and alcohol were among the most popular), how they were adjusted, and the scales they employed. Even with thermometers that could simultaneously measure temperature with multiple scales—W.E. Knowles Middleton (1966) notes, for example, two thermometers that had eighteen—there was still the difficulty of comparing measurements between thermometers made by different artists, let alone deciphering what the degrees of any single scale really meant. To standardise such measurement, there were efforts to identify phenomena that were *always* observed at the same temperature. These phenomena would provide thermometry with so-called 'fixed points'.

Natural philosophers largely came to agree that the boiling and freezing points of water should serve as *the* fixed points. Yet, as Chang (2004) shows, 'the consensus was neither complete nor unproblematic' (11). For example, Jean-Andre De Luc claimed that there were still 'uncertainties'. Boiling was less of a single phenomenon and more of a continuum of phenomena, he found, raising the question of which point along this continuum should be treated as a fixed point, as 'true ebullition'.[2] And so, by 1777, De Luc had joined a committee 'to consider of the best method of adjusting the fixed points of thermometers' (Cavendish et al. 1777). Organized by the Royal Society of London and led by Henry Cavendish, the committee opened their report by acknowledging what was, and what was not, in dispute:

It is universally agreed by all those who make and use Fahrenheit's thermometers, that the freezing point, or that point which the thermometer stands at when

[1] On the 'invention' of the thermometer, see Middleton (1966), 4. In considering the question of invention, Middleton makes a distinction between a thermoscope and a thermometer. A thermoscope is an instrument 'intended to give some visible indication of changes in its condition with respect to heat', while a thermometer is a 'thermoscope provided with a scale'. For a chronology of events in thermometry, see Chang (2004), xvii–xviii.
[2] Chang (2004) identifies at least six types of boiling in De Luc's writing: 'common boiling', 'hissing', 'bumping', 'explosion', 'fast evaporation', and 'bubbling' (20).

surrounded by ice or snow beginning to melt, is to be called 32°; and that the heat of boiling water is to be called 212°: but for want of further regulations concerning the manner in which this last point is to be adjusted, it is placed not less than two or three degrees higher on some thermometers, even of those made by our best artists, than on others.   (816)

The report continued with 'two principal causes of this difference' (816). For one, there was the height of the barometer. As others had observed, atmospheric pressure greatly influenced the boiling point. Additionally, because of the amount of mercury in the thermometer, it was possible to heat some of the substance more than the rest when dipping the thermometer into the boiling water. To these causes of the boiling point's variability the committee added yet another consideration, the depth of the water. According to their predictions, 'if the ball [of the thermometer] be immersed deep in the water, it will be surrounded by water which will be compressed by more than the weight of the atmosphere, and on that account will be rather hotter than it ought to be' (Cavendish et al. 1777, 817–18). The committee thus intended not only to standardise barometer readings but also to test a vessel previously designed by Cavendish (1776) (Figure 5.1) which would contain the boiling water as well as its *steam*. Such an apparatus, Cavendish had suggested, would allow the mercury in the tube of the thermometer to reach the same temperature as that in the ball.

Experiments followed, often with complex results. For example, the boiling point changed with the intensity of boiling but in ways that depended on an array of variables: where along the pot heat was applied, how much of the pot was heated, the size of the pot, and the length of the thermometer as well as the depth of the water. In trials involving a long thermometer and a deep pot, the thermometer 'stood, 66 of a degree higher when the water rose 15 inches above the ball, than when it was only three inches above the ball'. Even with this experiment, however, the committee expressed that they could not infer a 'constant rule' (Cavendish et al. 1777, 821). If the depth of the water increased or if the vessel were made larger perhaps the relationship between the depth of measurement and the boiling point would change.

The committee also tested Cavendish's idea about steam. 'Perhaps a still more convenient method of adjusting the boiling point would be not to immerse the ball in the water at all', Cavendish (1776) had proposed, 'but to expose it only to the steam, as thereby the trouble of keeping the water in the vessel to the right depth would be avoided' (380). Confirming this prediction, the committee 'very seldom found any sensible difference [in the height of the thermometer] when the water boiled fast or slow', and when they did,

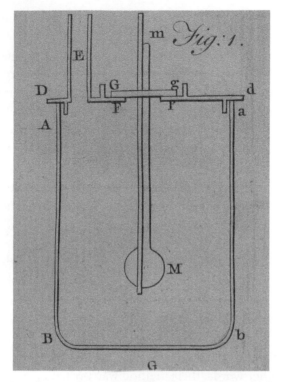

Figure 5.1 A drawing of Cavendish's vessel taken from the report of the Royal Society's committee

Notes: 'ABba is the pot containing the boiling water; Dd is the cover; E is a chimney for carrying off the steam; Mm is the thermometer fastened to a brass frame; this thermometer is passed through a hole Ff in the cover, and rests thereon by a circular brass plate Gg fastened to its frame, a piece of woollen cloth being placed between Gg and the cover, and the better to prevent the escape of the vapours'.

Source: Cavendish et al. (1777), 818–19. Courtesy of the Royal Society.

it 'never amounted to more than 1/20th of a degree' (Cavendish et al. 1777, 822–3). The temperature was also insensitive to where in the pot the thermometer was placed, the depth of the water, and the distance between the surface of the water and the top of the pot. Ironically enough, the temperature of boiling water was most stable when measurements were taken not of the water itself but of its steam.

The report concluded with three methods for using Cavendish's vessel, or at least one like it, to adjust the boiling point. We have summarised the 'rules' of the first method below (Cavendish et al. 1777, 845–50):

1. The 'operator' was to adjust the barometer to 29.8 inches (although they could also make their own corrections for atmospheric pressure).
2. To heat the mercury in the tube of the thermometer, the operator was to place the ball of the thermometer at the correct depth in the pot.

However, in doing so, they had to ensure the ball would not touch the boiling water if it were to rise.

3. To prevent air from entering the vessel and too much steam from escaping, the operator had to 'stop up' the hole through which the thermometer was inserted, as well as cover (but not block) the chimney.
4. To save time, the water was to be boiled 'briskly'. In addition, the 'observer' was to wait 'at least one or two minutes after the thermometer appears to be stationary, before he concludes that it has acquired its full height'.
5. Finally, the committee reiterated that all the rules be followed, and that the operator not deviate from them without first knowing how any change affected the entire experimental system.

Although this method was not the only one the committee proposed, in comparison to two others that involved measurements of the boiling water, it was deemed the 'most accurate'. As Chang (2004) writes, 'It gave a clear operational procedure that served well enough to define an empirically fixed point, though there was no agreed understanding of why steam coming off boiling water under a given pressure should have a fixed temperature' (27). In other words, their method was presented as a reliable one for recording the boiling point as 212 degrees Fahrenheit. But why? What made this scientific product (henceforth, M) reliable for what it was intended to do?

Since M was ruled-based, we might say that it was rigorous. Indeed, it seems to satisfy the formal features of rigour, as laid out in Section 2.2:

1. It's a nontrivial process
2. Which is auditable
3. And follows rules
4. Which are precise
5. And integrable
6. And there is a valid argument that shows that, if the assumptions are correct, the temperature should be measured as desired.

Yet, if M is a rigorous method, it is only because of a supporting body of thermometric knowledge and other scientific products without which M would make little sense. Why should the operator adjust the barometer to a standard atmospheric pressure? Why is it important to heat the mercury in the thermometer while also not letting the thermometer touch the boiling water? Why should the operator not let too much steam out of the vessel?

More generally, why adopt this method in the first place, and why is it preferable to the other two methods? The answers to these questions, and hence the perceived reliability of M, come from the virtuous tangle in which the method was embedded:

Rich. In Chapter 4, we propose that rich tangles consist of many different pieces of different kinds. In our view, the tangle the Royal Society's committee used to produce M was a rich one, as it was made up of a number of concepts, practices, technologies, and findings. It included not only the committee's new work—the predictions, experiments, and findings reviewed above—but also a great deal of past work in thermometry. The beginning of their report explicitly or implicitly references many of these previously constructed products:

- The concept of a fixed point
- The two-point system (the use of two rather than one fixed point)
- Fahrenheit's thermometers (those that contained mercury and used Fahrenheit's scale)
- The practice of using ice or melting snow to measure the freezing point
- Observations that Fahrenheit's thermometers consistently recorded the freezing point as 32°
- The practice of dipping Fahrenheit thermometers into water to measure boiling point
- Observations that Fahrenheit thermometers *often* recorded the boiling point as 212°
- Observations that Fahrenheit thermometers *sometimes* recorded the boiling point as greater than 212°
- Cavendish's recommendation about steam
- Cavendish's vessel.

These products are integral to the project of finding a fixed point—we cannot have M without them.

Entangled. These products of thermometry were entangled not only with M but with one another. Take, for example, observations of the boiling point. Whether these were of a fixed temperature or not, those that made these observations assumed that there were phenomena with such temperatures, that there were fixed points. Though De Luc raised doubts about the fixity of the boiling point, the concept of a fixed point was one that was already made reliable—for rationalising measurements of phenomena—thanks in part to work on the fixity of the freezing point (which would also come under question). Observations of the boiling point also depended on certain technologies, explanations, and practices, such as the substance mercury, ideas about why mercury functioned as a reliable thermometric substance, Fahrenheit's interval scale, and standard ways of using Fahrenheit thermometers to measure the boiling and freezing points of water (e.g. dipping the thermometer into boiling water). That these products of thermometry were highly interrelated should come as no surprise. As Chang (2004) argues, from the earliest times to the late eighteenth century, there was a gradual improvement of thermometric standards. This process was an iterative one, with each stage building upon the last in a self-correcting fashion.

Long-tailed. This tangle also had long tails, specifically into the related yet separable domain of barometry. Recall that the report identified the height of the barometer as one of the primary reasons for the apparent variability of the boiling point. In acknowledging this relationship between the boiling point and the height of the barometer, the committee reviewed the work of Fahrenheit and De Luc:

> Fahrenheit found that the heat of boiling water differed according to the height of the barometer; but supposed the difference to be much greater than it really is. Mr. De Luc has since, by a great number of experiments made at every different height above the level of the sea, found a rule by which the difference in the boiling point, answering to different heights of the barometer, is determined with great exactness. According to this rule the alteration of the boiling point by the variation of the barometer from 29 ½ to 30 ½ inches is 1°.59 of Fahrenheit.
>
> (Cavendish et al. 1777, 816–17)

Together, these long tails help to explain the first rule of M—that there should be a standard barometer reading, and if steam were used, it was to be 29.8 inches.

Gaps. As we discuss in Section 4.2, it is 'business as usual' in the sciences to evaluate the successes of its products by looking for gaps in the tangle and then filling them. This was true for the Royal Society's committee. Prior to the advent of M, the then standard way of recording the boiling point was simply to immerse the thermometer into the boiling water. When related to other pieces of thermometric knowledge and products, however, particularly observations of the variability of the boiling point, the committee identified what we may describe as a problem of entanglement. Although the boiling point was thought to function as a fixed point, measurements of its temperature varied. The concept of a fixed point did not 'fit' with observations from either Fahrenheit or De Luc. At this point, the tangle was rich and long-tailed, but its pieces were not entangled in the 'right' ways. While not couched in these terms, the committee was explicit from the start about the need for 'further regulations'.

Constraints. Here we can see a central mechanism for reliability in action, namely that of constraint. Moreover, we can see how the process of 'properly' entangling pieces of the tangle allows for greater constraints and hence for a greater push towards reliability, at least from science's point of view. In attempting to interrelate the products of thermometry such that the boiling point remained fixed, the committee imposed upon itself many constraints. Because of the relationship between the boiling point and atmospheric pressure, they would have to specify a barometric height at which the thermometer would record the boiling point as 212°. Because of the relationship between the boiling point and how the thermometer was placed into the water, they would have to either specify the depth at which the thermometer was to be immersed or avoid measurements of the boiling water altogether. Because of the relationship between the boiling point and the intensity of boiling, they would have to specify the level of boiling at which measurements were to be taken. And so on.

To not just build upon but also correct existing thermometric standards, Cavendish and his colleagues worked within a complex network of constraints, 'locking themselves in' as a result. In other words, to have produced a method *not* related to one or more of these thermometric relationships would mean to have produced *a less reliable method*, since it would have left out a factor known to shift the boiling point away from 212°. Indeed, the committee presents methods other than M as less 'accurate' precisely because of their use of boiling water. It was because of these heavy constraints that M was seen as the most reliable method.

This conclusion suggests a kind of circularity. The perceived reliability of M came from how well it fit with other scientific products. Yet this is exactly what should be expected if a tangle of background products underwrites the success of any single scientific product. Which products scientists can add to a tangle to support the reliability of a given product are conditioned not just by the purposes to which the product is to be put but also by the tangle to which they are adding them—not just by the three general features we have identified but by its specific pieces (including ones from other domains) and the ways in which these are interwoven (or not). For if a virtuous tangle supports the reliability of one of its pieces, then the other pieces of that tangle also constrain what that piece can look like.

5.2 The Making of an Antibiotic

In *Penicillin*, Robert Bud (2007) tells the story of perhaps the most histori-
cised drug of the twentieth century. In this section, we use and at times
elaborate upon parts of Bud's narrative to provide another in-depth illustration of a virtuous tangle, this time placing greater emphasis on long-
tailedness as a critical feature of tangles that support reliability. As Bud
himself writes: 'Penicillin was a biological product whose manufacture
would require the integration of a variety of scientific and engineering disciplines... Expertise in bacteria had to be linked to experience with moulds,
chemistry, and the engineering of sterile systems for the support of living
creatures' (23).

Counter to popular mythology, the discovery of penicillin was not simply
a matter of serendipity. Upon finding a mould growing in petri dishes left
aside in his laboratory at St Mary's Hospital in London, Fleming was already
well aware of lysozyme, a 'remarkable bacteriolytic element' he had discovered and named several years prior (Fleming 1922). First detected in the
nasal mucus of a patient with a cold, Fleming later confirmed its presence in
the fluids and tissues of humans, as well in rabbits, guinea pigs, dogs,
turnips, and egg whites. As Bud (2007) argues, 'through his experiments [on
lysozyme], Fleming developed a range of techniques for investigating the
effect of a chemical on bacteria that would stand him in good stead in the
future' (25).

In a preliminary set of experiments, Fleming placed a drop of concen-
trated mucus on a plate with which he had cultured the innocuous bacteria
M. lysodeikticus. 'Here there was a complete inhibition of growth, and this
inhibition extended for a distance of about 1 cm beyond the limits of the
mucus', wrote Fleming (1922, 307). In more complicated experiments, he
removed some agar from a typical culture plate and, in its place, added
human tears. They were then mixed with liquid agar and given time to set.
After adding a thin layer of agar onto the plate, he cultured *M. lysodeikticus*
on its surface. A day later, Fleming found that the entire surface was coated
in the bacteria, save for a region over which the lysozyme-rich material was
located. The substance, Fleming reasoned, had diffused through the agar and
into the bacteria on the surface. He concludes his 1922 paper on lysozyme by
stating:

> The view has been generally held that the function of tears, saliva, and sputum, so
> far as infections are concerned, was to rid the body of microbes by mechanically
> washing them away... From the experiments detailed above, however, it is quite

clear that these secretions, together with most of the tissues of the body, have the property of destroying microbes to a very high degree.   (317)

Fleming would adapt these techniques to his study of the newfound mould. After identifying it as a species of penicillium and testing thirteen known moulds for their antibacterial properties, Fleming did something similar to what he had done before: he filled a trench he had carved into an agar plate with a mixture of liquid agar and the broth of the mould, the latter being referred to as simply 'penicillin'. In this instance, however, he spread both pathogenic and non-pathogenic bacteria (including *M. lysodeikticus*) at right angles from the penicillin, tracking the growth of each microbe. 'This simple method therefore suffices to demonstrate the bacterio-inhibitory and bacteriolytic properties of the mould culture, and also by the extent of the area of inhibition gives some measure of the sensitiveness of the particular microbe tested' (Fleming 1929, 227). As it turned out, the substance only targeted certain bacteria, and to differing degrees.

This is not to say that Fleming only drew upon his prior work on lysozyme. In his 1929 paper, Fleming compares penicillin to both the bacterial extract pyocyanase, as characterised by the bacteriologist Rudolf Emmerich and biochemist Oscar Loew decades earlier, and the antiseptic carbolic acid (235–6). In addition to sharing certain chemical properties with penicillin, pyocyanase 'resembles penicillin also in that it acts only on certain microbes. It differs however in being relatively weak in its action and in acting on quite different types of bacteria'. As for the antiseptics,

> Penicillin, in regard to infections with sensitive microbes appears to have some advantages over the well-known chemical antiseptics. A good sample will completely inhibit staphylococci, Streptococcus pyogenes, and pneumococcus in a dilution of 1 in 800. It is therefore a more powerful inhibitory agent than is carbolic acid and it can be applied to an infected surface undiluted as it is non-irritant and non-toxic.

Beyond this knowledge of other antibacterial substances, lysozyme included, Fleming relied upon a panoply of already established technologies and methods. From the various bacteria that were cultured, to the Seitz filter (used to remove bacteria from a solution), to the agar plate method, Fleming wielded a tangle to which he added.

Of course, penicillin was not yet a medicine. Fleming's tangle may have been somewhat rich, but its pieces came primarily from just one domain, bacteriology. Perhaps not surprisingly, Fleming presented penicillin not as a reliable treatment for bacterial infections but as a reliable *tool* for bacteriology: 'In addition to its *possible use* in the treatment of bacterial infection penicillin is *certainly useful* to the bacteriologist for its power of inhibiting unwanted microbes in bacterial cultures so that penicillin insensitive bacteria can readily be isolated' (1929, 236, emphasis added). Fleming demonstrated this second aim using B. influenza, which he misidentified as a bacterium and 'cultured' alongside penicillin-sensitive staph.

For penicillin to become more than a laboratory technology, the active substance first had to be isolated. This was something Fleming, a bacteriologist and physician, did not achieve.[3] Neither did the chemist Harold Raistrick and his colleagues, whose method of isolation rendered the 'labile' substance inactive (Clutterbuck et al. 1932). During the 1930s, however, a multidisciplinary team led by the pathologist Howard Florey formed at Oxford University that would make headway here (Figure 5.2). 'With an interest in addressing the underlying causes of illness', writes Bud, 'Florey established strengths in clinical pathology, experimental pathology, and bacteriology, but focused on projects that crossed the divides between these specialities' (2007, 28). It was ultimately Ernst Boris Chain, a German biochemist and a member of Florey's group, who repurposed a freeze-drying method so as to successfully separate penicillin from the mould broth, though the yield was incredibly low and the chemical remained unpurified. Interpreted within our framework, the products of Chain's success as well as those of Raistrick's failure—their respective methods, findings, and materials—represented long tails extending from a tangle of bacteriology into one of biochemistry.

Later events gave rise to equally critical tails into pathology. To simplify this complicated history, in 1940 Chain collaborated with the surgeon Josep Trueta and the physician John Morrison Barnes to carry out the first test of penicillin's antibacterial properties in living animals. In mice injected with staph, only those that received penicillin went on to survive the infection (Trueta 1980, 150). These preliminary observations were reproduced and extended in a series of experiments by Chain, Florey, and others, published the same year as a brief report in *The Lancet*. Not only did penicillin appear physiologically safe but also effective as a treatment for mice infected with any of three pathogens, so long as the drug was administered over a long enough period of time (Chain et al. 1940). As *in vitro*, so *in vivo*.

In the subsequent year, Florey's team presented a much more detailed report on the still impure drug, including a discussion of 'its further development to the stage of human therapy' (Abraham et al. 1941, 177). In ten

---

[3] Fleming never published on this failure. He continued, however, to study the antibacterial properties of penicillin as a filtrate of the larger mould: see Fleming (1932).

Figure 5.2   A group portrait at Oxford including Howard Florey (back row, second to left) and Ernst Boris Chain (back row, second to right)

Source: Wellcome Library, London. Wellcome Images, images@wellcome.ac.uk; http://
wellcomeimages.org. Copyrighted work available under Creative Commons Attribution only licence
CC BY 4.0, http://creativecommons.org/licenses/by/4.0//.

different cases of staph and strep infections, with patients differing in age, gender, and the route by which they received the substance, all but two recovered. One patient, a middle-aged policeman, was given too small a dose and for too short a period, the authors suggested. He was also the sole patient to show signs of penicillin toxicity, attributed to the drug's impurities. The second, a four-year-old boy, died not because of the infection, which the drug had nearly cleared, but of an unexpected aneurysm. 'Enough evidence, we consider, has now been assembled to show that penicillin is a new and effective type of chemotherapeutic agent, and possesses some properties unknown in any antibacterial substance hitherto described', Florey and his colleagues concluded (1943, 188).

These trials were quickly followed by another, much larger set. They went beyond tests of efficacy and toxicity in human subjects to detail *methods of drug delivery* (Florey et al. 1943). If one wished to administer penicillin systemically, an injection into muscle seemed to be the most practical route. If instead one applied it locally, directly to the site of infection, they would need to do it frequently and until the infection resolved. Armed with such knowledge, Florey, along with the neurosurgeon Brigadier Cairns, went on 'to study the uses to which penicillin might be put in the treatment of war injuries' (Cairns et al. 1944). Acting as consultants for the British government during the Second World War, they oversaw the administration of penicillin to forty soldiers who had sustained head wounds and developed bacterial infections. Of those treated, thirty-five recovered.

We can summarise these developments by saying that, more than a decade after Fleming's initial work, penicillin was no longer just an object of bacteriological study. Successfully treating bacterial infections in a safe manner demanded a mix of materials, methods, experiments, case studies, and data from other disciplines. As we discuss in Section 4.6, to support the reliability of a scientific product for some aim, a tangle must constrain both that product and its aim. If an antibacterial chemical initially discovered as a product of a mould is to function as a medicine, then the tangle in which this chemical is to be embedded must be constrained by scientific products that are not just found in bacteriology. Put in the language of our analysis, it needs long tails that extend into medicine too.

We are not suggesting here that the history of penicillin was predetermined by this aim. If there was progress in the development of penicillin, as Bud and the authors of this book agree there was, then it was only made possible thanks to a dizzying number of political, economic, and social contingencies. Still, if we frame such progress as partly being about the aim of penicillin, as decided and pursued by certain historical actors, then we can see why virtuous tangles require long tails. As Florey remarked just a few years after his report on wounded soldiers, 'Nothing more was done [by Fleming] towards [penicillin's] introduction into medicine between the early 1930s and the beginning of the next decade .... After some preliminary work [at the Oxford laboratory] it became clear that a team of workers with special knowledge in various branches of pathology, biochemistry, and medicine would be needed for rapid progress' (1946, 164). Florey's team filled what they saw as 'gaps' in a tangle, relative to a specific aim, using long tails.

Because of the multidisciplinarity of this history, we can even go as far as to say that 'penicillin' was not a single scientific product but many. In Fleming's bacteriological work, it was mould broth. In Chain's biochemical work, it became a freeze-dried powder. Later, in therapeutic trials, it took the form of either a powder to apply locally or an alkaline solution to give systemically. Today penicillin is understood as a family of antibiotics whose members differ in their effective routes of administration.

Yet we wish to again emphasise that, however virtuous a tangle might be, it does not guarantee that a product of science will do what scientists expect of it. At best, it is only *conducive to* reliability. As we discuss in Section 4.5, science itself is only partly responsible for what is in a tangle. There are also constraints imposed by nature, which may or may not allow a scientific product to have its desired effect(s). For example, in Fleming's original study on the mould broth, not all bacteria were sensitive to its antibacterial substance. Among those that were sensitive were staph. In fact, Edward Abraham and Ernst Chain (1940) would ultimately use these bacteria, and specifically those of the species *S. aureus*, in control experiments to test the mechanism by which bacteria resist penicillin. And yet, in Florey and Cairns' wartime research, two patients with head wounds developed infections with penicillin-resistant *S. aureus* (Cairns et al. 1944, 121). Their tangle, rich, entangled, and long-tailed as it was, did not and indeed could not ensure that penicillin would always treat bacterial infections. Nature also had its point of view.

6

The Tangled Principle of the Democratic Peace

Preview

'Democracies don't go to war with other democracies.' This is the democratic peace principle (DPP), which we will discuss extensively in this chapter. This principle has been the subject of a huge body of research in political science over the last sixty years. It is widely assented to, but the reasons for it holding are heavily debated. As one distinguished political scientist reported to us, 'Everyone took it to be true, then jumped in trying to explain it.'[1]

As with, we hope, most of our chapters, this chapter has ideas to offer independent of the role it plays in developing and defending our overall claims about tangles and their role in supporting reliability in science. In particular, here we offer our own positive account of the DPP that we claim makes more reliable (although more restricted) predictions than any of the other accounts we have found and has behind it a better tangle to warrant that reliability.

With respect to our overall project to explain and defend the tangle, we use the extensive studies of the democratic peace (DP) to do four jobs.

First, our own positive contribution to the DP literature introduces new products into the tangle of support for the principle. As is often the case, this alters the product (in this case what the principle says) and refines the description of the job it can be expected to be reliable for. This illustrates the usefulness of the tangle, and of the focus we urge on a broad swathe of product types, as instruments for evaluating and contributing to ongoing scientific debate.

Second, in Section 4.3 we argue that tangles make failings less likely by supplying a network of constraints that makes it hard for something to go wrong. There we explore briefly how a particular measure of democracy, V-Dem, is constrained by a surrounding network of empirical and theoretical

---

[1] In conversation with Stephan Haggard, 25 February 2020.

The Tangle of Science: Reliability Beyond Method, Rigour, and Objectivity. Nancy Cartwright, Jeremy Hardie, Eleonora Montuschi, Matthew Soleiman, and Ann C. Thresher, Oxford University Press. © Nancy Cartwright, Jeremy Hardie, Eleonora Montuschi, Matthew Soleiman, and Ann C. Thresher 2022. DOI: 10.1093/oso/9780198866343.003.0007

research. In this chapter we use the work around the DP to illustrate how constraints help build tangles in contexts where there are simultaneously in play multiple competing understandings of what looks to be 'the same' principle and multiple competing theories about it. Although we illustrate with a social science example, the points we make hold equally in the natural and the social sciences. These constraints can push in opposing directions: they can drive divergence between rival theories or they can encourage them to converge on similar positions. In Section 6.3 we illustrate how the *Ballung* nature of the terms 'democracy' and 'war' pushes theories apart, expanding on Hempel's concept of alienation that we discuss in Section 3.2. Conversely, we show how the existence of standard touchstones for the field, as well as inter-theory critique, prevents theories from moving too far away from one another, as they are required to draw on similar resources to build their tangles or risk being discarded by the scientific community.

Third, Chapter 5 provides examples of tangles in different domains of the natural sciences. This chapter provides concrete examples of tangles in the social sciences and of what our three criteria for virtue look like in a real social science case. We are keen to look in detail at an example in social science because tangles in the social sciences may differ considerably from those in the natural sciences for all the standard reasons that explain why success in the social sciences is generally harder to obtain than in the natural sciences. These include a number of problems facing the social sciences that are particularly salient in our discussion of the democratic peace:

1. As Max Weber ([1904] 1949) argued, the social sciences are supposed to study concepts we care about. They are not at liberty to forsake these and shift focus to those that behave nicely. (Recall mention of this in Section 3.3.4.)
2. These concepts are very often difficult to define precisely and tend to manifest differently in different settings.
3. The things picked out by a great many social science concepts are socially constructed, like marriage, poverty, socioeconomic status, money.
4. So, their behaviour is governed not by natural law but by convention, legislation, and habit, and although any of these may promote certain behaviours they far from guarantee them.
5. The social sciences are generally expected to deal with complex open systems and are not able to shield them off in the way that so many of our successful physical creations are, as in the casing on an ordinary battery, the capsule that commonly encloses the drugs we swallow, or the Faraday cages that shield magnetic resonance imaging (MRI). As

the Norwegian economist Tyrgve Haavelmo, who won a Nobel Prize for his work in founding econometrics, notes: 'Physics has it easy. No one asks physics to predict the course of an avalanche. But economists are expected to predict the course of the economy.'[2]

6. Relatedly, as J.S. Mill ([1836] 1967) argues, by contrast, for instance, with the planetary system, there are apt to be a large number of factors at work at once and in shifting arrangements. This makes prediction extremely hard.

These special problems facing social science help explain why the notion of a *Ballung* concept is particularly important for the social sciences, why social science principles must so often be in the form of generics, and why overdetermination of outcomes is so central for reliable prediction there: to secure an effect with relative certainty, throw all the causes possible at it. These three—the *Ballung* character of 'democracy' and of 'war', the generic form of many social science principles, and the importance of 'overdetermining' outcomes—all play a central role in our own account of the DP, as we explain next.

Fourth, although we claim that our account is normative, we do not see ourselves as recommending that the sciences do anything different in general from what they already do. The tangle is always there in good science, although that has not, we think, been fully realised and articulated.

In particular, filling gaps in a tangle to make it more virtuous is business as usual in science. There are two ways to do this that we have only implicitly noted before. One is to propose something new that you think may do the job required. That can be a path to scientific advancement and discovery—better science in the future. But it is generally not the soundest way to fill a hole in a tangle for this product to do this job here and now. If this tangle is to provide support here and now for the reliability of this product, you need a filler that itself is warranted, here and now, as reliable for the job that needs doing in the tangle. So, generally for this task, it is better to take a product 'off the shelf'—of course, as always, with caution and requisite adjustment. This is one of the principal reasons it is important for scientific products to be clearly labelled, even if only implicitly: 'This product is attested as reliable for this job.' That is a central way in which science can claim to be cumulative and to build an ever-expanding store of new products efficiently on its store of already attested ones.

This chapter provides an illustration of how products developed elsewhere that weren't at first seen as relevant to the task at hand can be taken off the

---

[2] Personal conversation in Oslo, sometime in the 1990s.

shelf and, with appropriate adjustment, put to good use in new contexts. If successful, it is a poster child for wide-ranging interdisciplinarity. For in the case of our own development and defence of the reliability for prediction of a new version of the democratic peace principle, one of these shelves is labelled 'philosophy products'. In particular we use the three products that, we note above, help address the problems that make life difficult for the social sciences, two of which are philosophical contributions.

The first is our defence of the practice of retaining and employing loose *Ballung* concepts in science.[3] In this case the relevant concepts are 'democracy' and 'war' (though we focus primarily on democracy for illustration). We propose a special reading of the DPP itself and of the predictions it makes using *Ballung*-style concepts. We argue that the sixty or so years of intensive work on the DP provide good reason to predict that what count as 'democracies by anybody's book' will not engage with each other in ways that will count as 'war by anybody's book'. In proposing this, we suggest that the principle should be read as a principle defined only over core cases: the *(DPP) By Anybody's Book* ($DPP_{ByAnybodysBook}$). We explain this in Section 6.4.

Second, in analogy with our defence of loose concepts we also, throughout the book, illustrate the central role played in science by generic principles and the practices that the social sciences have evolved for their use, despite their inexact content.[4] This is how we read the standardly offered explanatory principles for the DP. We take these to be generic principles that pick out what Jon Elster calls 'mechanisms'—which often refer to dispositions of individuals, institutions, or structures in society. These are like the causal principles in the process-tracing theories of change in Section 2.7—the causes cited may need to be triggered, they generally need support factors to operate, they can be interfered with, and more than one can operate at once. This is not the usual reading of these principles in the DP literature, but we believe that it is the most plausible one, and it provides a very different take on the DPP than is otherwise available.

The third product we call on is a familiar tactic that is used successfully over and over again in practical technology and social policy: when you can't be certain that any one cause can be relied on to carry through from start to finish to produce the effect you need, introduce a number of different causes, the more the better so long as they do not interfere with each other. We use

---

[3] The distinction '*Ballung*' versus 'precise' concept is, we claim, a reliable way to classify concepts according to how significant context is in fixing what they amount to in a concrete setting.

[4] The category 'generic principle' is, we maintain, a reliable way to mark out principles that you wish to count as true or false even though they are not castable as proper quantified propositions, and that refer to natural or social mechanisms that may be present sometimes but not always in the targeted kinds of settings, that may need triggering, and that may be overwhelmed by the operation of other mechanisms.

this idea to build a kind of overdetermination argument for the reliability of the DPP.

In the case of the warring camps in high-temperature superconductivity (described in the Preface), the great volume of work that supports the explanatory mechanisms of each camp undermines the claims of those endorsed in others since it is assumed on all sides that only one of these explanatory mechanisms can be at work. So, too, with the explanations of gravitational waves that we talk about in the Afterword.

The opposite is the case with the explanatory mechanisms offered for the DP, as we read them. Each can operate under certain circumstances, and in many cases, the circumstances allow that many can operate at once. So, we shall argue, in 'bog-standard' cases peaceful outcomes will generally be over-determined by a number of different mechanisms all working to the same outcome. You can expect it to be unusual for all to suffer sufficient interference at once in cases where democracies by anybody's book come up against one another. We develop this argument in Section 6.5.

We begin in Section 6.1 by describing the bones of our argument and in Section 6.2 by briefly surveying the literature on the DP to provide context for what is to follow.

6.1 Introducing the Democratic Peace and Our Claims About It

In 1898, French and British troops found themselves at the centre of an international crisis focused on Fashoda, Sudan. Several dozen members of the French military, seeking to evict the British from Egypt, took possession of a fort in a region upstream of the Nile. Some months later they were confronted by a several-hundred-strong force of combined British and Sudanese troops. While interactions on the ground were largely amicable, back in Europe the incident became the locus for recent tensions between the two imperialist powers over control of Africa. The British public, in particular, agitated for military intervention, while the French were more reticent, mindful of a weaker armed presence at the fort and unwilling to risk alienating Britain with an aggressive Germany on their borders. After months of negotiation and with rising fears of wider conflict in Europe over the incident, the French finally ceded the fort to the British, ending the standoff and paving the way for British dominance in the region.

This result might be thought of as surprising. Prior to this, Britain and France had had a rough relationship characterised by both short and extended conflicts, and the two were currently locked in a land-race for territories in Northern Africa, lending pressure to defend their position aggressively. Given this, and despite worries about Germany, one might expect that such a confrontation would, as it often had before, lead to open conflict between the two nations.

What made the difference? Political scholars argue that it was the newly extended democratic nature of both. That is, the Fashoda Incident, as it has come to be known, is seen as part of a broader pattern in international politics where democracies do not go to war with one another. This is the central principle of the DP. It is as close as political science gets to a foundational rule. Empirically, most academics agree that the principle holds and continues to hold across almost all modern and historic international conflicts. There are, as of today, no generally agreed-upon instances of generally agreed upon democracies going to generally agreed upon war with each other.[5] Democracies don't go to war, and so one could have reliably predicted that the Fashoda Incident would not end any other way.

The DP has been the backbone of many areas of political and international policy research for decades. It helps set agendas, drive treaties, and has been cited by numerous American presidents as central to their international politics (e.g. Wilson 1917; Bush 2014). Despite this broad-scale acceptance and use, however, the means by which the DP instantiates remains controversial. Numerous explanations have been proposed, but no one has emerged as a clear winner.

Proponents of these explanations have long argued that their particular theory is the 'correct' explanation for why the principle holds. We think this is a mistake.

Our take on the DPP involves three theses.

1. Many of the different theories on offer about the DPP have what look to be reasonably virtuous tangles to back them up. So there is good reason to expect that they reliably identify explanatory mechanisms responsible for the DPP.
2. The different credible theories converge on what we call $DPP_{ByAnybodysBook}$. The tangles behind these theories are all entangled with one another in multiple ways. This tangle of tangles supports the reliability of the predictions about peace and war that follow from $DPP_{ByAnybodysBook}$.

---

[5] Here we should make an important note about the scope of the DPP. It *does not* hold that democracies are never in conflict, nor that the two do not act aggressively towards one another. As the recent history of Western, and particularly American, behaviour in the Middle East and South America can attest, democracies routinely interfere with, and do harm to, one another. The DPP instead claims that democracies do not declare *open war* with one another. Covert operations are, however, still on the table. It also operates under modern definitions of democracy—it is hard to argue that Greek democracies never went to war with one another (though not impossible; see Russett 1994, ch. 3).

As part of this, one of the primary goals of this chapter is to expand on the constraint work done in Section 4.3. We single out three kinds of constraints that do much of the heavy work in support of reliability for the individual theories and the DPP_{ByAnybodysBook}.

I. The kinds of constraints demanded of good science, like consistency, looking for and responding to empirical evidence, and ensuring concept validity, to name a few. These force theories of the DP to adopt precise definitions of the concepts they employ, beginning with democracy and war and building out from there. We illustrate how these constraints reinforce divergences among the different theories. We call constraints *divergence constraints* when they function this way.

II. In contrast to this, we claim there are two types of *convergence constraints* at play which entangle the theories together. These take the form of

   A. *Responsiveness constraints*, generated by the demand that each theory be responsive to the standards of the field.

   B. *Competitive constraints*. As we review in discussing HD-with-bells-on in Section 4.1, it is not enough for a theory to provide evidence for its own hypothesis. It should also show what's wrong with alternatives. Each theoretical camp is watching for problems in the alternatives and working to guard against any that competitors might identify in their own explanations.

III. Finally, we conclude that the different theories are not in competition after all. The explanations they identify can all be 'correct'. This is because the explanations use not universal but *generic* principles that pick out Elster-style mechanisms, and these can all operate at once.

The existence of these divergent and convergent constraints helps us see why the DPP is reliable for making predictions about conflicts. While each theory predicts its own particular version of the principle, all agree on what we term the DPP_{ByAnybodysBook}. Contrary to standard robustness arguments, however, we do not take this convergence to suggest that there must exist some single 'correct' explanation for the DPP. In contrast to what is commonly supposed in the literature, no individual theory is, in itself, enough to ensure that democracies don't go to war with each other. Instead, each of the wealth of theories, constrained by and interwoven with their respective tangles, provides one possible Elster mechanism for the DPP, described in a defeasible *generic* or *ceteris paribus* law. Taken together, they ensure that democracies by anybody's book don't go to war by anybody's book. In many specific conflicts, various of the proposed mechanisms may fail to trigger, be shielded off, or otherwise be absent, but the chance that *all* of them fail is small. So the DPP is not a case where political scientists should be looking for the 'one true explanation'. It is a case where the combined weight of the theories is, itself, the reason the DPP can be expected to hold—at least between bog-standard democracies with respect to bog-standard wars.

Our argument is, as we note in Section 4.6, reminiscent of Woody's (2015) functional analysis of science, arguing as she does that science is 'a coordinated activity of communities' (5). Convergence of the kind discussed here may be seen as one part of her broader coherence work on how science is reliant on scientists working as a group to constrain divergence.

6.2 Theories of the Democratic Peace

There are a great number of highly developed theories of the DPP on offer, some compatible, others conflicting—each with its own tangle of support. It will help going forward to briefly review a number of these proposed theories. Following Harald Müller and Jonas Wolff (2006), we roughly categorise them along two dimensions—the monadic/dyadic divide and the normative/structural divide.[6]

The monadic/dyadic dimension attempts to capture something about how democracies might interact with one another. A *monadic* theory is one which holds that there is something internal to democracies which makes them less prone to enter into wars, an effect which is amplified when their opponent also possesses these particular traits. Conversely, a theory may take there to be something critical to *dyads* where both states are democracies. Under this type of theory, democracies are just as warlike as their non-democratic counterparts, but some new element comes into play when they face off against another democracy.

The normative/structural divide attempts to capture a difference in what theories think is the driving force behind the peace. *Normative* explanations take the role of democratic institutions to be transmissive in nature—they allow normative preferences for peace and conflict-avoidance to guide national policy, but the structures themselves have no role beyond this transmission. *Structural* explanations, conversely, take the institutional features of democracies to be central to the success of the DPP. Here, these structures play a larger role, often in a capacity that forces one or both parties involved

---

[6] Much of this section, including the distinctions and categorisations of theories, is drawn from Müller and Wolff (2006).

Table 6.1 Types of DPP explanation

	Normative	Structural
Monadic	Kant's Perpetual Peace	Increased Signalling
Dyadic	Social Constructivism	Democratic Commitment

Source: Ann C. Thresher.

in a conflict to back down or delay, allowing time for conflict to be resolved, defused, forgotten about, or otherwise avoided.

Table 6.1 highlights four current DPP theories which demonstrate the monadic/dyadic and normative/structural divide. This is not, of course, anywhere near a complete listing.

6.2.1 Kant's Perpetual Peace [Monadic-Normative]

Immanuel Kant is a monadic normative theorist. In his book *Perpetual peace: a philosophical text* ([1795] 2020) he suggests:

> [T]he Republican Constitution...includes also the prospect of realizing the desired object: Perpetual Peace among the nations. And the reason of this may be stated as follows. According to the Republican Constitution, the consent of the citizens...is required to determine at any time the question, 'Whether there shall be war or not?' Hence, nothing is more natural than that they should be very loath to enter upon so undesirable an undertaking; for in decreeing it, they would necessarily be resolving to bring upon themselves all the horrors of War....On the other hand, in a Constitution where the Subject is not a voting member of the State...the resolution to go to war is a matter of the smallest concern in the world.   (Section II, Part I)

Ernst-Otto Czempiel recasts this in the following way:

> The rational citizen in liberal capitalist societies is generally peace prone because war endangers not only his life (as combatant or civilian victim), but is economically expensive as well. If the political system allows for the translation of this preference into foreign policy (like democracy does), the respective state will refrain from violent behaviour.   (Cited in Müller and Wolff 2006, 80)

This can be seen as a normative theory because the relevant factor is the preferences of the citizens; the democratic institutions merely allow for preference transmission. It is monadic because democracies are supposed to have this tendency towards any war, regardless of their opponent. When both sides are reticent, war is avoided. In mixed dyads, however, the non-democracy acts as the instigator and can force the democracy into war against the citizenry's preferences.

A related theory, with similar structure but slightly different content, can be found in the 'cultural argument'. Müller and Wolff characterise this particular normative/monadic theory as the idea that

> Democracies (democratically socialised citizens and leaders) are used to and prefer to solve their conflicts in peaceful and consensus-oriented ways. In democratic societies prevails a "democratic norm of bounded competition".   (9)

That is, whereas Kant takes it that all citizens wish for peace but only democracies make this relevant on the inter-state level, one might alternatively think that citizens of democracies are socialised to be less conflict-prone.

6.2.2 Social Constructivism [Dyadic-Normative]

Thomas Risse-Kappen (1995) is the progenitor of social constructivism, arguing that

> [Democracies] to a large degree create their enemies and friends—'them' and 'us'—by inferring either aggressive or defensive motives from the domestic structure of their counterparts. On the one hand, they follow behavioural norms externalizing their internal compromise-oriented and non-violent decision rules in their interactions with other democracies. On the other hand, the presumption of potential enmity creates a realist world of anarchy when democratic states interact with authoritarian regimes.   (1)

Because democracies follow internal norms of peaceful conflict resolution using compromise and negotiation, Risse-Kappen suggests that the DP is the result of democratic states assuming similar norms for other democracies. Thus, they are more likely to trust overtures and negotiate in good faith, assuming the other party is similarly peaceful. In mixed dyads, however, the democracy has no such belief, instead often seeing autocracies and the like as overly aggressive and untrustworthy, exacerbating tensions.

This, in turn, tends to lead to clusters of democracies working together along several dimensions including trade and military agreements. This further reduces the chances that these states engage in open war with one another. Conversely, such groupings tend to exclude non-democracies lead-ing to a prevalence of tension between democratic and non-democratic states.

6.2.3 Increased Signalling Theory [Monadic-Structural]

This monadic-structural explanation for the DPP draws on the idea that states are rational but lack perfect information. In the absence of complete information about their opponent's desires, commitments, and capabilities they are unable to accurately call bluffs, judge stakes, or trust overtures. This, in turn, leads them to engage in war with other states when peaceful moves are disbelieved or bluffs are misread as more aggressive than intended.

Under this assumption, states which are better at signalling their motiv-ations and capabilities are less likely to be drawn into wars with other states. The *Increased Signalling* theorist, then, suggests that because of how democ-racies are internally structured they signal more openly and accurately than other types of government, leading them to engage in war less often. This property, in turn, is boosted in democratic dyads where both states are clearly signalling to one another.

The types of structures that are central to this theory vary from proponent to proponent, but the generally offered ones include the existence of elections which are determined by the public. Elected officials, wishing to stay in office, are under pressure to conform with general public opinion—a widely available piece of information for other states due to the existence of free speech and public debate in democracies. Thus, if the leaders of a democracy signal in line with public opinion, they are more readily believed, while signals which go against the public are more often accurately called as a bluff. Similarly, the Increased Signalling theorist might point to the existence of oppositions as a means of increased information—governments which are backed by their opposition are less likely to be bluffing, governments which are opposed by them are likely to not be as strong as they wish their rivals to believe.

6.2.4 Democratic Commitment [Dyadic-Structural]

The democratic commitment theory rests on the idea that leaders in democ-racies face harsh penalties if they enter into, and then lose, a war in that they are likely to be replaced by the public in the next election. Conversely, lead-ers of autocracies need only keep a small subset of supporters satisfied, something that can be achieved via various, often easily accessible, means. Thus, democratic leaders are less likely to enter into a war unless they're certain to win, and once engaged they are likely to commit more fully than their autocratic rivals.

Democracies are, then, more difficult opponents to provoke and more deadly once war has been entered. In mixed dyads, this manifests as a slightly increased chance of war, as democratic leaders know they have a better chance of winning against autocracies which will put less effort into the fight. Conversely, when two democracies butt against one another, the leaders of both sides are aware that their opponent would be more difficult to defeat, leading to a trend away from war as the states are wary to start something unless they are certain of victory.

6.3 Defining Democracy and War: Fuzzy and Precise Boundaries

As we argue in Chapter 4, in order to have a credible claim that they provide a correct explanation of the DP, each of these theories must be supported by its own rich, tangled, and long-tailed network of definitions, measures, principles, theories, models, observations (amongst other things). They must use sound principles in sound ways with sound, validated concepts.

Central to our work in this chapter, however, is that each theory has a *different* underpinning tangle, though the different tangles share much in common. They may draw on different measures of public opinion, emphasise different structures of state government, or require the existence of different means of information transmission between states and citizens, to name a few.

In this section, we focus on constraints that enforce this divergence. In particular, we look at the role that definitions of 'democracy' and 'war' have in constraining a DP theory. These terms are, as they come, *Ballung* concepts, fuzzy at the edges and comprising a set of overlapping but related defining features. Their fuzzy nature is one significant source of theory-tangle divergence, with each theory attempting to precisify the concepts along different lines. As we will show, however, divergence can only go so far, since each definition is beholden to the same standard candles of the field, which we term *responsiveness constraints*.

6.3.1 Divergent Definitions and Measures of Democracy

Setting aside 'war' for the moment, let us focus on 'democracy', which is so widespread a concept that one would be hard-put to find many academics who agreed entirely on how it is to be defined and measured. Attempts range from that found in Frederick Whelan's 1983 book,[7]

> ...[G]overnment with the consent of the governed. This formula is indeterminate with respect to institutional forms, or the procedures by which consent is to be expressed – questions on which consent theorists have historically differed.   (15)

to Elmer Schattschneider (1960),

> ...[A] competitive political system in which competing leaders and organizations define the alternatives of public policy in such a way that the public can participate in the decision-making process.   (141)

to Robert Dahl (1991), who provides a more formal set of criteria whereby for a state to count as a democracy, its citizens must have

1. Effective participation
2. Voting equality at the decisive stage
3. Enlightened understanding
4. Control of the agenda
5. Inclusiveness (109–15).

We have also already talked about the variety of measures put forth by political science, including V-Dem, which we discuss extensively in Section 4.3, including the various constraints and elements that go into the tangle that supports their reliability as accurate measures of democracy. Alternatives to V-Dem include Freedom House, which takes into consideration two axes along which states are scored—political rights and civil liberties; Vanhanen's Index, which ranks nations on competition and participation and provides an aggregate score; and Polity IV, which uses weighted scores for a number of categories including 'Competitiveness of Executive Recruitment', 'Constraint on Chief Executive', and 'Regulation of Participation'.

These discrepancies in definition and measures are unsurprising when one considers that democracy, much like the terms 'poverty', 'objective', or 'chair', fails to have a clean set of defining features, but instead is, to paraphrase US Supreme Court Justice Potter Stewart (1964) on pornography, the kind of thing where 'you know it when you see it'.

---

[7] Note that this isn't a definition Whelan himself endorses, but one he recognises as a common, if underspecified, attempt at defining democracy. We have used his wording because it is a nice succinct definition.

6.3.2 Divergence Constraints: Making 'Democracy' Precise

The formulation of a proper scientific explanation for the DP, however, requires that theorists give a clear definition of what they mean by 'democracy'. It is not enough to gesture at the *Ballung* concept and claim a theory applies over its instances since an explanation is necessarily predicated on particular democratic properties of the state. This predication, in turn, generates a subset of the *Ballung* democracy concept for which the theory supposedly holds, thus doing violence to the original fuzzy concept by moving from the vague to the precise. Recall Hempel from Section 3.2.2, 'An explication sentence does not simply exhibit the commonly accepted meaning of the concept under study but rather proposes a specified new and precise meaning for it' (1952, 663). What we see here is a form of alienation, with all the benefits and downsides we point out in our discussion of objectivity.

Take, for example, Kant ([1795] 2020), who emphasises the role of constitutional structures in democracy which enshrine equality:

> The only constitution which derives from the idea of the original compact, and on which all juridical legislation of a people must be based, is the republican. This constitution is established, firstly, by principles of the freedom of the members of a society (as men); secondly, by principles of dependence of all upon a single common legislation (as subjects); and, thirdly, by the law of their equality (as citizens).   (Section II)

This definition fails to capture aspects that other theorists take to be central to a democracy. In contrast to Kant, the Increased Signalling theorists' definition of democracy must involve the existence of elections, a free press, or open debate because it is upon these structures that their proposed explanation hangs, while the Social Constructivist needs to consider democracies to be those states with internal norms of peaceful conflict resolution. Including states which lack this property means creating a theory which necessarily cannot hold over the objects it is defined across.

Indeed, each DP theory will require some different function from the concept 'democracy' and so, while some may share a definition, many will diverge in the precisification they adopt. Note, however, that by moving to a precise definition of democracy, the concept necessarily loses information. What was once a fuzzy set of related concepts is now rigidly defined along one or more specific dimension. Though moving from a *Ballung* concept to a precise definition is a requirement of a functional theory, it necessarily involves excluding states which other people may wish to count as democratic, thus forcing divergence between theories.

6.3.3 Convergence Constraints: How Uncontroversial Examples Constrain the Tangle

Each theory has a different meaning for the term 'democracy', brought about by divergence constraints demanding precisification. But no DP theory can use a definition which differs too greatly from the core concept or it risks being discarded. This gives rise to *convergence constraints*. These come in two flavours:

A. *Responsiveness constraints*, which include,
  i. The existence of uncontroversial examples of democracies and non-democracies, which is the focus of this section.
  ii. The existence of extensive literatures about democracy, as Crasnow stresses (recall Section 4.5).
  iii. The existence of diverse measures that give roughly correlated results over various ranges, as we illustrate in Sections 4.4, 4.5, and 6.3.1.
  iv. The existence of respected databases that record which states have which of a variety of different features associated with democracy, as for example V-dem, Polity IV, and Freedom House.
B. *Competitive constraints:* The existence of critique from other theorists, which we return to in some depth in Section 6.3.4. Note that competitive constraints can also be a force for divergence as theories attempt to distance themselves from problems they perceive in other theories.

Here we focus on the first responsiveness constraint, the existence of uncontroversial examples of democracies and non-democracies, which plays a critical part in our argument for why the DPP is reliable for making predictions, drawing each theory inwards and forcing them to converge on a single, overarching, formulation of the DPP.

Uncontroversial examples are what we term 'democracies by anybody's book'. Democracies by anybody's books are examples of democracy which must be covered by any definition of the term offered by any theory or that theory risks having an unforgivable hole in its tangle. Examples of democracies by anybody's books include Australia, the United Kingdom, and the United States of America. Any definition of democracy which fails to include these, by any theorist's light, has made a mistake and so it, and the DPP theory which draws on it, is generally discarded from the mainstream body of science. Consider, for example, a definition of democracy which claims that only states in the northern hemisphere can be democratic, excluding democracies by anybody's books like South Africa and New Zealand. It would immediately be thrown out by the political science community.

The existence of a set of pieces to which all tangles must be responsive is something found in every field. Scientists can and should discard any theory of quantum mechanics which doesn't fit with the results of the double-slit experiment, for example, or an explanation for fish evolution patterns that doesn't consider the salmon a fish. To ignore these central pieces is to have a bad tangle because one is missing critical, agreed-upon standards for the field. Again, we see the scientific community constraining divergence.[8]

A similar argument may be run on the definition of 'war'. Each theory of the DPP will attempt to precisify the *Ballung* concept, losing fuzzy cases in the process and creating disagreement over how the concept is used. Just as with democracy, there will also be a set of wars that are 'war by anybody's book' (WBAB) (e.g. First and Second World War, the Vietnam War, and the first Sino-Japanese War), and sets of conflicts that become points of contention for their inclusion or exclusion under the term (e.g. the Cold War, the Irish Troubles, or the existence of large-scale, persistent cyber-attacks between major world powers).

We should note that just as there are democracies by anybody's books and wars by anybody's books, so too will there be not-democracies by anybody's books and not-wars by anybody's books. It is just as bad for a definition or measure of democracy to decide that North Korea is a democratic state, or that the 'Scallop Wars' of 2012 and 2018 between French and British fishermen counts as genuine inter-state war. Going forward we focus on the positive inclusion restrictions, but readers should bear in mind that this is only half the story.

You can visualise the varying definitions of democracy and war as Venn diagrams, with the core examples that all theories must include in their definitions sitting in the centre (Figure 6.1).

---

[8] This is not to say that reasonable theories which reject the stable examples cannot both exist and do well. To make such a move, however, requires a significant tangle of work behind it to justify why one should disregard the established norms of the field. It requires a richer, longer-tailed, and more entangled tangle to make up for, and justify, the significant gap that comes with disregarding a standard example. Such instances are unusual, but they do occur in science. Take, for example, the creation in biology of the fungi kingdom, which moved mushrooms, at the time an example of a plant by anybody's book, out of the taxonomic grouping of plants and into the newly discovered category. Similarly, recalling from Section 3.2.1, reclassification of the term 'planet' moved Pluto—up until then a clear-cut case of a planet as far as most people were concerned—to the new 'dwarf planet' grouping. In both cases, overwhelming evidence was presented that the current systems were mistaken about the core examples which justified the rejection of 'by anybody's book' elements.

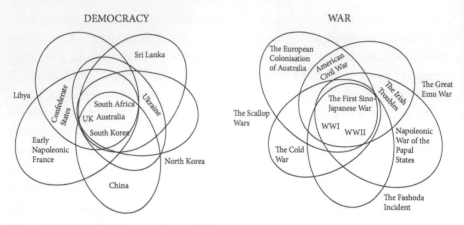

Figure 6.1 Overlapping concepts of democracy and war
Source: Ann C. Thresher.

6.3.4 Inter-Theory Criticism, Fringe Cases, and Constraining the Tangle

Once one has established the subset of elements which must be included or excluded by a measure or definition of democracy or war, inter-theory criticism (competitive constraints) drives how the other responsiveness constraints are applied. There are two ways this occurs.

1. The field as a whole will reject any theory which fails to adequately meet the responsiveness constraints. This acts to excise theories which wander too far from the rest along the given dimension.
2. Each theory, due to including and excluding a different subset of tangle pieces from other theories and other domains, will criticise other theories for failing to meet what it judges to be appropriate standards.

The second way rests on the existence of fringe cases. Fringe cases, in this context, are those states which may or not be democracies depending on how you look at them. This directly leads to inter-theoretic critique: because Theory A includes 1849 France as a democracy while Theory B does not, the two will act to constrain each other, pushing and pulling the other theory to justify its inclusion or exclusion.

Such conflict is inevitable. While precise definitions cannot wander far from the core set of democracies by anybody's books, the very act of precisification will create edge disagreements as some theories include or exclude states that others take to be clear democracies or clear non-democracies by their own light.

Why think France was a democracy in the period leading up to the outbreak of the War of the Roman Republic between Napoleon and the Papal states? Recall Whelan's (1983) 'government with the consent of the governed' definition from Section 6.3.1. Napoleon, the leader at the time, had the support of the masses and was duly elected leader of France, and so it seems it must count as a democracy. Alternatively, Schattschneider (1960) required that 'the public can participate in the decision-making process'. By this light, France was not a democracy—Napoleon infamously declared himself emperor shortly after this period, indicating a trend away from public participation and towards dictatorship, and France at the time lacked any other politically competitive groups besides Napoleon himself (Ray 1993).

Which is the correct decision? Was early Napoleonic France a democracy or not? Not only is it unclear which is right, it's unclear that such an absolute result even exists. Both Whelan and Schattschneider advocates, however, will act to force the other to justify their choice of definition.

Let us consider another example, the American Civil War as discussed by James Ray (1993). Some academics have suggested that this conflict constitutes a violation of the DPP. A lot hinges both on whether each was a democracy, and whether the conflict between them counts as an international war between two states as opposed to an internal conflict.

On the first point, both had almost identical constitutions, suggesting that if the North (as is standardly thought) was a democracy, so too was the South. As has been pointed out by academics like Ray, however, while the South's leader, Jefferson Davis, was technically elected by representatives of each state, there had not yet been open elections such that it could be said he had the mandate of the broader citizenry. His position, moreover, was supposedly a temporary one, but war broke out before a general election could be held.

If one thinks, as the Democratic Commitment theorist does, that what defines a democracy is the ability to be removed from control, then the South looks like a democracy.

Alternatively, if one requires that instances of peaceful norms for conflict resolutions be demonstrated, the short-lived nature of the state might not be enough to make it a democracy under the Social Constructivist position, so the conflict is not a failure of the DPP.

There are, of course, good reasons to think that both the first definition and the second are equally valid conceptions of democracy. The first prioritises legal structures, while the second attempts to exclude those states which pay lip-service to democracy while remaining in practice a non-democratic system.

If a theory wishes to use a definition of democracy that includes the Confederacy, but doesn't want to say that the American Civil War was a violation of the DPP, then they will need to carefully choose their definition of war to reflect this, either by excluding the Confederate states from being a genuinely independent state, or by gerrymandering their definition of war so that the conflict counts as an internal civil war rather than an open war between two nations. This gerrymandering may mean an even larger discrepancy between theories, however. As one contorts a certain way to avoid violations to the DPP, another will contort in a different direction, further limiting the types of wars that the theory is taken to be able to predict outcomes over and creating even stronger fringe-case divergence.[9]

These divergences, then, are a driving force for why DP theory tangles are conducive to the reliability of the theory to provide an accurate explanation of why the DPP holds. Critique binds the tangles behind different theories together by forcing them to interact in dialogue with one another, creating a virtuous tangle of rich, entangled connections between the different theories. These connections, in turn, make each theory more reliable because the field as a whole is better at culling bad apples. Thus, if a theory's tangle is sufficiently responsive to the standard pieces, and it engages with other theories regularly without being completely discarded from consideration, one should think that it is more reliable than theories which fail to pass these bars.

Of course, this goes far deeper than democracy and war. This argument applies to many of the other pieces that go into making up a 'good' theory of the DP. For each measure, definition, observation, study, dataset, mechanism, or other piece, there will be similar constraints and norms that it must adhere to or risk a large, non-virtuous hole in its tangle and thus also risk being excised from the field.

6.4 The Reliability of the Democratic Peace by Anybody's Book

We have spent a lot of time on the pieces that make up a DP theory and how they are constrained, but what does all this mean for the reliability of the DPP as a whole? In this section we will see how a responsiveness to

---

[9] As a brief aside, there are reasons why one should be wary when one sees this type of contortion to fit evidence. It is reminiscent of the Dead Sea Scroll example from Section 4.5, which we argue is an example of a bad tangle. Recall that Wylie maintains that while the Qumran-Essene hypothesis is both rich and tangled, the tangle lacks long tails. Each piece turns inwards to support its own internal consistency and story rather than drawing on broader claims, observations, and theories from archaeology, anthropology, theology, geology, and more. Similarly, the twisting that needs to happen to include or exclude controversial examples of the DP will potentially reduce the long-tailed nature of the tangle as the theories start to pull further away from broader work in other fields on democracy and war, looking inwards and lacking strong external support.

democracies by anybody's book and war by anybody's book, and inter-theoretic critique, leads to the existence of a core DPP, termed the $DPP_{ByAnybodysBook}$. This, we suggest, is the DPP that has proved so useful in political science, not some supposed correct, precise formulation which has so long been the goal of DPP theorists.

As we've just seen, each theory of the DP will have its own particular for-mulation of democracy and war which it is predicated over. This, in turn, means that each proposed theory of the DP will have its own unique version of the DPP. Thus, one no longer has 'The Democratic Peace Principle', but instead there is DPP_{Kant}, $DPP_{SocialConstructivism}$, $DPP_{IncreasedSignalling}$, and so forth, each one supported by a distinct tangle and providing a distinct account of why the DPP holds. Moreover, these different formulations make different predictions, some of which are incompatible with each other. Each posits rules that cover a different set of conflicts over different dyads of states. They can make predictions—whether correctly or not—only over these instances and not over all the instances covered by other formulations.

As with our myriad of overlapping definitions for democracy and war, however, while each theory is distinct there will also be significant overlap caused by the existence of democracies by anybody's book and war by any-body's book. Every theory will, necessarily, admit to the following principle:

Democratic Peace Principle By Anybody's Book ($DPP_{ByAnybodysBook}$): democracies by anybody's book don't go to war by anybody's book with other democracies by anybody's book.

This picks out the centre of the Venn diagram of DP theories (Figure 6.2).

Now one can begin to see more clearly why, even with the proliferation of possible DPP formulations and fundamental disagreements between them, the DPP remains reliable about a great many predictions. The DPP is not one particular formulation which happens to be 'correct', but it is better thought of as a principle formulated across democracies by anybody's book and war by anybody's book, and thus amenable to explanation by a host of different theories.

Let us return to the Fashoda Incident. Why could you reliably predict that it wouldn't end in war between France and Britain? Because both countries at the time were democracies by anybody's books. Very much in brief, according to the Perpetual Peace Principle, both Britain and France were countries where leaders were in some way dependent on the general public to hold their seats, Britain having passed the Representation of the People Act and Redistribution of Seats Act in 1884 and 1885, respectively, which

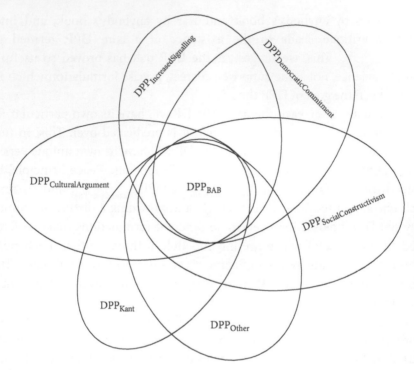

Figure 6.2 Overlapping DPP theories
Source: Ann C. Thresher.

increased the voting population to approximately 60 per cent of all men and France being firmly in the middle of the Third Republic which had universal male suffrage to elect the chamber of deputies which, in turn, dictated the ministries. Each had a common legislation (dictated by Parliament in the UK and the Chamber of Deputies in France), freedom (provided one was a man with land), and applied the law equally to all citizens (at least officially). Thus, according to the Perpetual Peace theorist, neither country would go to war by anybody's book with the other.

For the Social Constructivist, both France and Britain were systems which used internal norms of peaceful conflict resolution, including parliamentary debate and voting. Thus France, when presented with the possibility of war with Britain in tandem with threats from the non-democratic Germany on its border, was more likely to infer from their own norms that British overtures for peace could be trusted more than German ones and to seek alliances along those lines.

Increased Signalling theorists simply require that there be open and externally accessible debate within both countries, which is given by the existence of rival parties sharing a common legislative floor in both states.

And finally, the Democratic Commitment theorists take both France and Britain to be democracies for the purposes of their theory because, as with Kant's Perpetual Peace, both countries had leaders who were beholden to the general public for their jobs, placing them in precarious positions if they entered into wars they would lose. Indeed, the French were explicitly wary of entering into the conflict because they feared the superior might of the British military and recognised that further instability on a military front would pose a significant threat to the current government which was already embroiled in the Dreyfus affair. Leaders were hesitant to commit unless they were guaranteed victory and knew the other was in a similar position.

Thus, for each of the four kinds of theories discussed in Section 6.2, the Fashoda Incident sits in the centre of the Venn diagram—the part that matters for $DPP_{ByAnybodysBook}$—even though each emphasises different aspects of the states in maintaining their theory.

Before going further it will be useful to recognise how what we have just said differs from standard robustness arguments. Robustness, as generally held in the literature, is 'the idea is that if there are many ways of measuring, detecting, producing or deriving something, and if those ways are sufficiently independent, then it is very unlikely that all of them turn out to be mistaken or erroneous' (Eronen 2019, Section IV).

There are two standard versions of this, one where the input pieces are compatible and one where they are incompatible.

Begin with the compatible. This kind of robustness supposes that there are a number of different but compatible arguments with different credible premises that yield the same conclusion. You may allow that no one of these arguments has premises that are entirely certain. Yet, by hypothesis, the premises are all supposed to be reasonably credible—they have a lot to back them up. Suppose, for instance, that you have two relatively established compatible theories, and three different kinds of experiments calling on different techniques and assumptions that have been carefully conducted as well. All lead to the same conclusion. It seems you can generally assume that it is very unlikely that all of these, all of which point to the same result, are flawed at once, a rare coincidence you are usually justified in discounting.

Where the input arguments are incompatible, the defence of the conclusion rests on the assumption that at least one of the many ways of getting to it is right. Again, it is supposed that the set of arguments is comprehensive enough to make it unlikely that all are flawed.

Our case involves a set of theories which almost all converge on $DPP_{ByAnybodysBook}$. Note that if you suppose that all of the arguments for a given result are flawed, neither form of robustness gives reason to accept that result. So if you are to accept $DPP_{ByAnybodysBook}$ (DPP_{BAB}) on the grounds of robustness, no matter which version you choose, you are left having to assume that at least one precise version of the explanation for the DPP is correct, even though the political science community has not yet settled on which it is.

This is why we do not employ a standard robustness argument. We think that there is good reason to take the predictions of DPP_{BAB} to be reliable. But we also think there is good reason to suppose that none of the proffered explanations are 'the correct one'. So our defence of the reliability of DPP_{BAB} must rest on different grounds.

6.5 Overdetermination and the *Ceteris Paribus* Nature of Democratic Peace Principle Theories

Instead of robustness, we want to suggest that because claims about why the DPP holds only have meaning relative to a tangle of ingredients that give the DPP and the claims meaning, including but not limited to the definitions of democracy and war, you can't simply appeal to the idea that one explanation is the 'true' one.

There are two reasons for this. First, because no matter how good each precise theory is, it is predicated on the precisification of a *Ballung* concept and thus will necessarily fail to capture the broad principle in full. This, as we've already discussed in depth, is one reason that each theory is exposed to criticism by other competing theories.

When one precise theory critiques another, it is often by observing legitimate failures and weaknesses in the other's inability to fully capture everything the field as a whole wants to count as an instance of the DP. No single tangle is reliable for supporting a version of the DPP that consistently predicts the outcome of conflict between democratic dyads that are not in the limited subset it is predicated over. For instance, DPP_{Kant} is only relevant for predictions about war_{Kant} between $democracies_{Kant}$, and no others, and for every theory of the DPP that isn't $DPP_{ByAnybodysBook}$ there will exist examples that others cover successfully that they can't.

Second, because you should not read the principles in these theories as if they had universal quantifiers in front. None of the features they select out as characteristic of democracy are strong enough to compel peaceful resolutions. Recall our discussion of process theories of change in Section 2.7. There we argue that a causal process can only carry through if all of the requisite support factors are in place at each step and no interferences strong enough to halt the process occur. For many DPP theories, the tangles behind them look strong enough to support the claim that they are conducive to peace. But none has enough to expect that the support factors will always be there when needed nor to expect shields strong enough to prevent interferences from ever stopping that outcome.

Because of this, we urge that the different causal principles that the theories endorse are best seen as generics or *ceteris paribus* principles, which pick out what we call 'Elster mechanisms', after Jon Elster (2015). In the case of the DP, these mechanisms will generally be social, political, cultural, and psychological dispositions. They may need triggering. And even when properly triggered, they do not compel the associated outcomes but are only conducive to them via what are usually long causal chains. If other mechanisms are simultaneously at work pushing in other directions, their effects can be diminished, distorted, or totally overwhelmed.

This is why the principles that describe them should generally be expressed as generics or have the label *ceteris paribus* affixed at their start. John Stuart Mill ([1836] 1967) calls them 'tendency' principles because they tell not what effect a feature will produce but what effect it *tends* to produce.

Let us take a moment to look at one of these proposed theories—Kant's perpetual peace—as an illustration (Figure 6.3).

Here we have a (rough) process theory of change for Kant's perpetual peace. Let's focus on the connection between 'Citizens do not want to put themselves through the horrors of war' and 'Citizens are against war'. Under ideal circumstances for Kant this connection holds, but as a principle it relies on a specific connection: war means citizens go through horrors. One can imagine, however, numerous scenarios where the citizenry do not expect this connection to hold. It might be that the war would be fought overseas and with little impact on the average citizen (as e.g. with many of America's modern wars in the Middle East).

The connection holds *ceteris paribus*. This is also true of other steps in the perpetual peace process theory of change. Leaders may have overriding reasons to go to war without their citizens' consent (perhaps they have information the general public do not, or it may be the case that the leaders are not up for election for several years, in which case they believe they can repair their reputation in the intervening time), or citizens may give permission to go to war even knowing that there will be horrors visited upon them (perhaps because of overruling nationalistic tendencies in citizens who are willing to bear the burden in the name of pride or resources). The perpetual peace is not an absolute law, but a tendency in those states with the republican structures picked out by Kant's theory. Naturally, these tendencies may

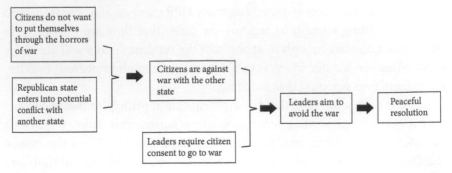

Figure 6.3 A rudimentary theory of change for Kant's perpetual peace
Source: Ann C. Thresher.

be overruled, or specific necessary aspects may fail to trigger given certain circumstances.

All DPP theories will have this property of employing *ceteris paribus* or generic principles. Each invokes a tendency of those democratic systems they are formulated over which can be overruled or otherwise derailed by other factors. There's no reason to think that any of them holds in all situations or is indefeasible.

Because of this, each explanatory mechanism can be countered or otherwise fail to trigger. But, if as we claim, the DPP is formulated across democracies by anybody's book and war by anybody's book, then all the proposed theories will include democracies by anybody's book and war by anybody's book in their predictions. So, in each case, the mechanisms cited by every theory will simultaneously be viable means by which the DPP$_{ByAnybodysBook}$ can hold. DPP$_{ByAnybodysBook}$ would fail if all the theories turn out to be completely wrong despite the virtuous-seeming tangles that back them up, and/or all the mechanisms cited by them that are genuinely conducive to peace are disrupted or fail to trigger simultaneously. This is the coincidence that we think is unlikely. There's been a vast amount of diligent work on the DP since the early 2000s, creating a great tangle of tangles with long tails into other successful domains. This doesn't guarantee reliability, as we have noted (and will see in some depth in the Afterword), but it is good grounds for expecting it (at least from science's point of view). Thus, we argue, you can rely on DPP$_{ByAnybodysBook}$ to make accurate predictions. Not because any particular one theory is *the* correct one, but because it is overdetermined that at least one mechanism will be in place and work as predicted to prevent democracies by anybody's book dyads from going to war by anybody's book.

In sum, if what we have argued in this chapter is true, then the broad projects pursued by most DP theorists of searching for a single, true explanation is misguided. Instead of the theories being at odds with each other, the principle is reliable for bog-standard democracies and bog-standard wars precisely *because* there are so many plausible theories at play.

This is in contrast to other examples where multiple theories about the same phenomenon are genuinely in competition, such as the alternative mechanisms for high-temperature superconductivity discussed in the Preface and the gravity wave case we discuss in the Afterword, where at most one theory is expected to ultimately win out. If only one or a couple of the mechanisms cited by the different theories were genuinely conducive to peaceful resolution, you should expect exceptions to be more common even for bog-standard cases since so much else needs to be in place for the mechanism to produce the outcome it pushes towards. The DPP would be far less reliable.

This result is due to the types of constraints that are at play in the tangles of the DP theories, in particular to the tension between divergence constraints, which arise due to the fuzzy, *Ballung* nature of the concepts the principle is predicated over, and convergence constraints in the form of responsiveness to field standards and inter-theory critique. Reliability is secured by the fact that the differing theories are forced to agree on a single, overarching principle, $DPP_{ByAnybodysBook^*}$.

A Cautionary Lesson: The Study of Gravitational Waves

Preview

Despite the arguments we have made in favour of tangles throughout this book, we end with a caution. We want to emphasise and reiterate that we in no way claim to have solved the problem of induction; a virtuous tangle is conducive to reliability, but it is no guarantee.

To make this point we illustrate with a stark example of a controversy at the heart of what is generally taken to be physics' most fundamental theory—the general theory of relativity (GTR). This controversy centres on the results coming out of the Laser Interferometer Gravitational-Wave Observatory (LIGO), part of a huge international effort that aims to measure the effects of gravity waves.

Gravitational waves are on the cutting edge of physics and are intimately bound up in the web of formulae that define GTR. First proposed by Henri Poincaré in 1906, their existence was rederived by Albert Einstein eleven years later using his new theory of relativity. What followed was a long and tumultuous period of debate within the physics community about whether gravitational waves exist and whether they can carry energy, involving many of the most important scientists of modern history, culminating in the construction of the billion-pound experiment at LIGO.

Two different explanations of the results observed at LIGO are on offer, backed by two different tangles. One tangle is connected with what we call *High Church* theory and methods, the other with *Low Church*. High Church follows the mathematics and insists that gravitational waves cannot carry energy. Low Church assembles a host of more empirically based physics to argue instead that they do. Both of these tangles look to be exceedingly virtuous—rich, entangled, and long-tailed, each seems to offer solid support for the reliability of its respective account to provide a correct explanation of the results. But they cannot both be reliable for this aim since the explanations contradict each other. So there is no doubt that at least one of these very

The Tangle of Science: Reliability Beyond Method, Rigour, and Objectivity. Nancy Cartwright, Jeremy Hardie, Eleonora Montuschi, Matthew Soleiman, and Ann C. Thresher, Oxford University Press. © Nancy Cartwright, Jeremy Hardie, Eleonora Montuschi, Matthew Soleiman, and Ann C. Thresher 2022. DOI: 10.1093/oso/9780198866343.003.0008

virtuous-looking tangles cannot guarantee that its explanation is reliable for giving a true explanation of what is happening at LIGO.

The background of this debate is the subject of Sections A.1–A.4, covering the measurements at LIGO and the tangles behind the High and Low Church explanations, demonstrating how both are virtuously built out of differing pieces of 'well-established' physics that have proven reliable for a vast variety of different purposes. In Section A.5 we return to our caution, including a brief exploration of conflict in science, examining it through the lens of the gravitational wave case.

We should note that much of the controversy arises from criteria accepted within GTR for what counts as 'objectively real'. We steer clear of this issue. For our purposes it does not matter whether gravity waves 'really' carry energy, nor how you answer the question of which explanation of the results at LIGO is true, or whether either is, or more broadly whether science can ever hope to find that out. We want a clear case where, beyond doubt, a tangle that looks entirely virtuous with respect to a given aim nevertheless is not reliable for achieving that aim. There are very many examples that do this, but we want a clincher—beyond any doubt. For that, the aim of giving a true explanation just fits the bill. No one will deny that two contradictory accounts cannot both be true at once, so one or the other tangle cannot be reliable for explaining the results correctly.

A.1 How LIGO Detects Gravitational Waves

In brief, a gravitational wave is an oscillation in space-time generated by mass in motion. Think ripples in a pond when you dip your toe in and move it about. Unlike their water counterparts, however, gravitational waves are incredibly small relative to humans; while you and I send them out into space every time we move, detecting them would be like finding a single molecule of water by eye while looking at the earth from the moon. Unsurprisingly, then, it took over a hundred years between their initial proposal and their detection at LIGO in 2016 (Abbott et al. 2016). Because they're so small, the ones scientists have detected have come from massive systems, often black holes or binary stars orbiting one another and throwing off massive gravitational effects. Even so, the deviations in space-time detected at LIGO are over a thousand times smaller than the radius of a neutron.

LIGO, then, is a marvel of modern science and engineering. It is the culmination of over a hundred years of research and theoretical advances in

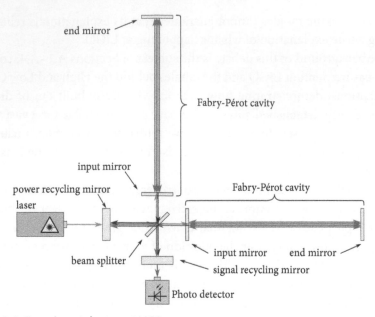

Figure A.1 Experimental setup at LIGO
Source: Wikimedia Commons (https://commons.wikimedia.org/wiki/File:LIGO_simplified.svg).

general relativity, resting on a complex tangle of skills, practices, measures, methods, and data, amongst other things, that make it a reliable tool for detecting gravitational waves. Put briefly, LIGO is at its core a large interferometer—a cross-like precise configuration of mirrors and lasers. As you can see in Figure A.1, the system is comprised of a single laser emitting a beam of light which, in turn, passes through a beam splitter that sends half the light in one direction and half in another. These two beams are each sent to a distant end mirror which in turn reflects them back to the beam splitter. The beams then recombine at the splitter and are sent to a detector.

The point of this system is to detect tiny movements in the end mirrors. Gravitational waves, passing across the system, will cause the mirrors to shift slightly as they interact with each of them in turn. This minutely changes the distances the laser beams travel before being recombined and sent to the detector. At the detector the scientists are looking at the distinct interference patterns the recombined beams create. This pattern changes depending on how far each of the two beams travel. If both paths change by the same amount, the pattern stays the same, if one beam or the other travels a longer or shorter distance, the pattern is different. These changes are then cross-referenced to predictions about the expected size and timing of pattern shifts one would expect from a gravitational wave, and when the empirical and theoretical data match, that is taken to be a detection event.

Now, obviously as a whole, the system is much more complex than this. Extraordinary work has to be done to prevent outside influences beyond gravitational waves moving the mirrors, and even then more work must be done mathematically to remove the effects that inevitably leak through. Indeed, the immense amount of work that goes into making LIGO reliable is worthy of a book all of its own and in and of itself is a remarkable example of how the tangle is put to work in science. Here, however, we are going to narrow our focus down to just a single component of the system: the end mirrors, whose tiny movements are what allow us to detect gravitational waves.

What exactly causes this motion is the source of intense debate within the physics and philosophy communities. It is an argument that goes all the way back to the beginning of general relativity when Einstein first proposed the existence of gravitational waves. Over the subsequent century two distinct schools of thought—what we call the High and Low Church approaches—have emerged, each with their own explanation for why the mirrors move. Moreover, as you will see, each explanation is underpinned by a rich, entangled, and long-tailed tangle that seemingly then justifies thinking that each is reliable. However, the two are in direct conflict with one another and cannot both be reliable for the same job. So even when you have something that looks overwhelmingly good you still need to be cautious.

A.2 High Tangles

Einstein prevaricated wildly on the topic of gravitational waves throughout his academic career. Initially he was dubious about the reality of the phenomenon. In a letter written to Schwartzschild in 1916 he states that 'Thus there are no gravitational waves analogous to light waves' (Schulmann [1916] 1998, 196).

That same year, however, he published 'Approximate integration of the field equations of gravity' where he proposes six types of gravitational waves, *a* through *f*, and states that

[T]here is a simple way to clarify the strange result that gravitational waves (types a, b, c), which transport no energy, could exist. The reason is that they are not "real" waves but rather "apparent" waves, initiated by the use of a system of reference whose origin of coordinates is subject to wavelike jitters. . . . only waves of the last named type do transport energy.   (Einstein 1997, 209)

Why such an emphasis on energy? As with ripples in water, gravitational waves should at least abstractly have some energy that was given to them by their source and which they take with them as they travel out into the universe. A water wave might, for example, push a leaf around on the surface or cause a bell to ring on a buoy. In both cases, the water is giving energy to the other item, doing work on it, while losing energy itself. Similarly, one might think that gravitational waves should perform a similar process as they interact with objects in the universe. Einstein noted, however, that if you actually look at what mathematical theory says, gravitational waves don't have any *observer-independent* energy associated with them.

To see why this fact is important, and where the current conflict comes from, we briefly need to dip into the theory of general relativity and how it deals with matter and energy.

Relativity is so named because it ties most phenomena to a frame of reference—an observer. How far have you thrown the ball? How fast is the ball moving? How long did it take for the ball to hit the ground? Einstein realised that to answer these questions you first needed to answer a different one: who's asking? Depending on how fast an observer is moving relative to the frame in question, objects may seem longer or shorter, faster or slower, or to take more or less time.

A natural question that arises once you accept that so much is relative is: what constitutes 'real' facts about the world? If the mug sitting on my desk is both 10 cm tall and 5 cm tall depending on who is looking (or rather, what frame they are in), how tall is it really? Einstein argued that the question was nonsensical, as the answer is entirely dependent on the observer's frame and there are no such facts about the mug. He also maintained, however, that underpinning these relative facts were some steadfast ones—information that stayed the same regardless of who was looking. While the mug may change size, mathematically it still has the same underpinning *stress-energy tensor*. What this means in a technical sense is too complex to talk through here, but suffice to say that this roughly boils down to the idea that while the size, mass, and shape of the mug may change, the energy distribution associated with it remains the same for everyone.

Of more interest to our discussion is the 'tensor' part of 'stress-energy tensor'. A tensor is a mathematical structure used heavily in relativity because, as physicists discovered, anything which could be represented mathematically as a tensor within relativity would have the property of being the same for everyone regardless of their frame of reference.

Many physicists take what tensors represent to be the building-blocks for the universe—features which ground our relative observations. You and I might disagree about how tall my mug is, but we both agree on the stress-energy tensor that represents it.

Conversely, this implies the inverse view held by the physics and philosophy community that if something does change depending on who's looking, then it cannot be a building-block. Marc Lange (2002), for example, says

> If two frames from which the universe can be accurately described disagree on a certain matter, then that matter cannot be an objective fact.   (207)

Similarly, Tim Maudlin (2011):

> The objective features of the world must be represented by invariant quantities. If many different reference frames are all equally valid, then the only statements which can be held as objective descriptions of the world are statements which are agreed upon by all the frames.   (32)

What Einstein and his contemporaries realised, however, is that one cannot construct a tensor which represents the energy of gravitational waves in the way one could for the mug or the water ripples. Instead, the formula which determines their energy is what physicists term a 'pseudo-tensor'— something that resembles a tensor but only under very specific circumstances, circumstances which don't generally hold in our universe. What many physicists at the time took this to mean was that gravitational waves, if they did exist, couldn't have any energy associated with them. It would be meaningless, they argued, to claim that the energy is real if I can simply make it disappear by speeding up or slowing down. Conversely, my mug may change height depending on who's looking at it—there's no frame where it simply goes away.

Einstein was well aware of this problem, writing in 1918:

> The values of the energy definitely depend upon the choice of coordinates, a fact Herr G. Nordström already pointed out to me in a letter some time ago. If the choice of coordinates is made with the condition $|g| = 1, \ldots$ then all energy components of the gravitational field vanish.   (1997, 215)

And,

> Together with a young collaborator, I arrived at the interesting result that gravitational waves do not exist, though they had been assumed a certainty to the first approximation.   (1997, 215)

Years later, Felix Pirani (1957) wrote along these lines that

The investigation of gravitational radiation in general relativity theory is hampered by the lack of an invariant definition of that concept. The presence of gravitational radiation must be distinguishable, mathematically, from a peculiar choice of the coordinate system, and physically, from a peculiar motion of the observer.   (1)

Let's take a moment to review. Gravitational waves are a highly theoretical phenomenon that is derived mathematically using the laws of relativity. These laws, in turn, tell you that the phenomenon you've derived must have certain properties: it moves at the speed of light, is emitted by massive objects in motion, and does not have an invariant energy associated with it. Thus, if one trusts the mathematics (and one generally does with relativity—widely taken to be among the most well-confirmed theories of physics) gravitational wave structures show up in the maths but these structures do not have an objective, underpinning energy that they carry with them away from their source, judged by the standards adopted in GTR for determining what is real.

This is the position we have labelled the High Church one. It leans into the tangle of science that has produced the mathematical structures of relativity, taken as one of the best confirmed and most reliable theories science has ever formulated. It allows us to put people on the moon, use mobile phones, and explore the universe through telescopes. The mathematical structures themselves are impressively well-confirmed and arise from a dense, entangled, and long-tailed network of pieces that make them reliable for describing how relativistic objects work. When the claim that there is no such thing as gravitational wave energy is grounded in relativity, pieces of the tangle behind it include

- Observations of gravitational lensing around black holes
- Accurate predictions about the 1919 eclipse as observed by Arthur Eddington
- GPS positioning systems
- The conceptual mathematical structures of tensors
- Concepts of space-time that underpin current best theories of gravity
- Measured results of the acceleration due to gravity of a ball dropped on earth
- Many many many more pieces found throughout physics and engineering.

It is *rich*, formed as it is from numerous diverse sets of pieces; *entangled*, as many of these pieces themselves are connected together—measurements of gravitational acceleration are entangled with theories of space-time curvature which predict local gravitational strength which allows the derivation of theoretical gravitational lensing effects which in turn are supported by observations from the field of astronomy; and *long-tailed*, drawing as it does from a set of equations which are used to build everything from GPS positioning systems to rocket ships and telescopes, as well as work by philosophers and physicists on ontology and space-time structures.

A.3 Low Tangles

In the mid-1950s a pair of meetings were organised in Bern, Switzerland (1955) and Chapel Hill, North Carolina (1957). These conferences were attended by many of the big names in relativity, including Nathan Rosen, John Archibald Wheeler, Ivor Robinson, Richard Feynman, and Felix Pirani.

At these conferences, Feynman (anonymously) proposed a thought-experiment that looked at gravitational wave energy from a new direction. This demonstration has come to be known as the 'sticky-bead' argument. It goes as follows. Imagine two beads strung onto a rigid rod. The beads are free to slide along the rod but are currently stationary relative to one another. A gravitational wave now passes across the bead–rod system, moving orthogonal to the rod's length (Figure A.2). Such waves, incident on the system in this way, will expand and contract space along the direction of the rod's length (recall that gravitational waves are ripples in space-time itself, which is why space expands and contracts as they move by).

What happens to the rod and beads experiencing these contractions? The initial force of the wave will try to move them outwards and then inwards. The rod, however, is fixed in length, being rigid and held together by interatomic forces. As such, it is prevented from expanding with the wave. Conversely, the beads are free to slide, and so while the rod stays the same, the beads will move along the rod under the influence of the wave. This, Feynman points out, will create friction, which creates other forms of energy like heat and sound. Where, he asks, does this energy come from? It cannot be the initially stationary rod and bead system, unless one thinks it is spontaneously releasing energy that just happens to coincide with the presence of gravitational waves. Instead, the most natural explanation is that the waves are doing work on the sticky-bead system, exerting a force that moves energy from the waves to the bead and then converts it to things like thermal energy via friction. In short, if you're getting extra energy out, the only place it could have come from was the waves which must, therefore, be energetic.

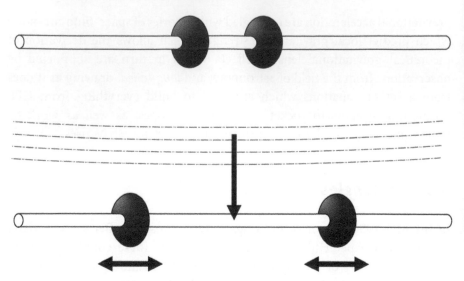

Figure A.2 Gravitational waves incident on a stick with two beads
Source: Ann C. Thresher.

This is the Low Church position. It relies on a different tangle of facts, measurements, definitions, and principles to come to its conclusion than the ones used by the High Church. In particular, it relies on an understanding not just of relativity, with all the experiments, measures, and structures that make it a reliable source to draw on for this kind of conclusion, but also on a wide range of other areas of physics including thermodynamics, kinetics, atomic theory, and energy conservation, making it particularly long-tailed as a tangle. Note that each of these are, as they come, often thought of as 'lower' physical theories—theories which are of practical use but, as with Newton's physics being replaced by Einstein's, are rougher in nature and narrower in scope than the 'higher' theories like relativity and quantum mechanics.

Moreover, the sticky-bead argument can be recast in a number of other ways. One might, for example, show that you can get work out of gravitational waves using springs and weights (Düerr 2019), or radar signals (Creighton and Anderson 2012) which yield more long-tailed arguments that connect to other well-entangled areas of physics.

Ultimately, we find ourselves with two conflicting claims; one approach says that gravitational waves do not carry energy, the other claims they must. Both approaches are backed by tangles, rich and long-tailed, but in very different ways. The first is more theory-oriented, resting upon reliable theoretical results from relativity and more mathematical approaches to space-time structures. The second is more pragmatic, connecting instead to broader areas of physics that are often considered less 'fundamental' but have a long-proven history of reliable use for the purposes we put them to.

A.4 Conflicting Explanations of What Was Observed at LIGO

Now that we've laid out the High and Low Church approaches, we can return to the movement of the end mirrors at LIGO. Both positions agree on what one should expect to see in the interferometer—gravitational waves causing the mirrors to shift slightly as the ripples pass over the system. The explanations for why they move, however, are vastly different.

For the Low Church position it is easy to see how one might draw a direct line to the sticky-bead argument. The mirrors are in motion because the waves themselves are transferring energy to them in much the same way that a buoy is rocked by waves on the water or our eardrums vibrate in response to sound waves. If you wished to, much like with the sticky-bead system, you could draw work from the mirrors, generating heat, sound, and light via friction.

In contrast, High Church explains the motion of the mirrors not in terms of the waves 'pushing' them, but in terms of the expansion and contraction of space-time itself as the waves pass over the system. The mirrors are shifted relative to one another because the underpinning space-time distances are themselves changing. As Wu (2016) notes, 'it [is] a pure geometric measurement' (730).[1]

So, the Low Church approach might yield something like this:

The reason that gravitational waves move the end mirrors is that the waves transfer energy from themselves to the mirrors.

Alternatively, you might get this out of a High Churcher:

The reason that gravitational waves move the end mirrors is that the waves stretch and contract the underpinning space-time within which the mirrors are situated.

---

[1] One might worry that the High Church will still need to explain why we seem to be able to draw energy from the system if there is no energy in the waves. One proposed explanation can be found in Düerr (2019) where he suggests that such systems are an example of the non-conservation of energy in general relativity, which has long been acknowledged as an implication of the theory under certain circumstances (Hoefer 2000). The waves, put roughly, do not transfer energy but instead generate it in the mirror system as they pass across it due to the stretching of space-time itself.

Now, this is a massive oversimplification of the debate, but it does capture the essence of the High and Low Church approaches and, importantly, we can now see how the two directly conflict.

A.5 Reliability, Conflict, and Our Final Caution

So let's talk about conflict. To see exactly what's going on in the gravitational wave case and why it means we can't assume that a product of science is reliable because it is supported by a virtuous tangle, we need to be a little more specific about how conflict can manifest in the context of tangles. The first thing to note is that conflicts can be *external* or *internal* to a tangle.

An internal conflict is one which arises within a single tangle when two pieces being drawn on to support the reliability of the same product for the same purpose are incoherent. The Low Church approach is a nice example of one such conflict. In talking about gravity waves in their explanation of the results at LIGO, Low Church scientists are drawing on the derivation of gravitational waves from general relativity. This derivation, however, is the very one which contains the problematic pseudo-tensor. As such, the Low Church tangle inescapably contains within it pieces which seem to contradict its explanation for why the mirrors move.

Internal conflict is sometimes problematic. We would be suspicious of a theory of the democratic peace which was supported by a tangle that contained both the claim that China is a democracy and the claim that China is not a democracy, for example. But in many cases, internal conflict is taken to be benign. In the Low Church case, the surrounding tangle is virtuous enough and draws on a wide enough array of pieces that Low Churchers take the associated explanation to be reliable for revealing the cause of the mirror's movement, despite the internal conflict. Indeed, these types of internal conflicts are common throughout science, which often works with approximations, rough measures, and fuzzy definitions that generate contradictions, even within tangles that look to very strongly support the reliability of various products that indeed turn out to be reliable.

On the other hand, we have external conflict. External conflicts arise between tangles rather than within them. They occur when two scientific products with the same purpose in the same context are each backed by a virtuous tangle, but both products cannot be simultaneously reliable for that purpose.

Here is a simple example. Scientists once thought that heat was the result of a fluid called 'caloric' that flowed from hot things to cold things. The theory was backed by a scientific tangle that drew on empirical and theoretical factors to justify why we should think of heat in this manner. This caloric theory proved to be reliable for numerous projects including providing a correct prediction that air expands when heated and the development of the Carnot cycle, which is still in use today. Conversely, the mechanical theory of heat supposes that heat transfer is due to the motion of particles moving energy between themselves and their less energetic neighbours. It, too, is backed by a virtuous scientific tangle, and it is the dominant theory in modern physics.

These two theories are externally conflicting in the following context. If I rub two pieces of metal together, I create heat. Caloric theory says that this heat is due to the caloric fluid being exuded from the metal. It thus predicts that eventually the metal will stop heating up once we have used up all the caloric fluid. Conversely, the mechanical theory of heat predicts that one can create heat as long as you have metal left to rub together. So, when I ask whether I can rely on the prediction to hold that I will always be able to produce heat by rubbing two pieces of metal together, the two theories give incompatible answers.

Note the importance of incompatibility here. It is possible for two tangles to produce two different products that are both reliable for the same purpose. Newtonian mechanics and general relativity will often give you different predictions about the downwards acceleration of a specific particle due to gravity, both backed by virtuous tangles. Though the predictions differ they can both be reliable for telling you what you will see within a certain margin of error if you measure the acceleration when these predictions lie close enough to each other relative to the error margin. External conflicts, as we define them here, occur where two virtuous tangles support the reliability of different products for the same purpose but both cannot be reliable for that purpose at once. Perhaps most straightforwardly, this occurs when the two products claim X and not-X, respectively. They thus cannot both be reliable for telling us whether X really holds or not because they cannot both be true.

The gravitational wave case is an example of an external conflict, one that arises between two products of science, each supported by different tangles. Thus, when we say that you can rely on the Low Church tangle to support a true explanation of the LIGO results and that you can rely on the High Church tangle to do the same, we are putting into conversation two different rich, entangled, and long-tailed networks backing two different products which give incompatible explanations of the LIGO results. What, then, can we learn from this? That at least one of them cannot be relied upon to give us the true explanation for why the mirrors move.

This, then, is a clear example of the caution we wish to leave you with. *A virtuous tangle is conducive to reliability, but it does not secure it.* Having a virtuous tangle is helpful, and it goes a long way in justifying you in taking a particular practice, measure, theory, and so on, to be reliable. But having a rich, entangled, and long-tailed network backing it up does not guarantee that a scientific product is reliable. The gravitational-wave energy case is a clear example, but it is only one of many. Some external conflicts have been resolved, as with caloric theory and mechanical heat transfer, where the tangle backing one eventually became so rich and so long-tailed into other highly virtuous tangles that it clearly dominated that of the other. Others remain, as with the conflict between general relativity and quantum mechanics, both of which involve extraordinarily virtuous tangles, which make contradictory predictions about phenomena like black holes and simultaneity that cannot both be relied on to hold.

Despite what we take to be the strengths of the tangle approach to science, then, it is not a panacea that straightforwardly secures reliability. As with many other parts of science, its application will require judgement and care.

A.6 Parting Thoughts

The Jacana bird is renowned for its floating nests. Here we have drawn extensive analogies between the methodologies and structures that allow the bird to construct such a reliable resting place for its eggs, and the methodologies and structures of science which allow us to rely on scientific products. Our account goes beyond what is usually on offer, which focuses on the scientific method, objectivity, and rigour. These usual suspects, as we argue here, cannot be the whole story. These indeed are virtues. But they are the kinds of virtues displayed by individual enterprises in science. Focusing exclusively on them is like looking only at the features of the individual elements that make up the Jacana bird nest, missing the importance of their variety, the way they entangle with one another and the way they entangle with distant reeds, with plant stems, and with floating vegetation. Each rigorous, objective enterprise of science conducted according to proper scientific method is part of a greater structure, woven out of a vast variety of other scientific products that includes everything from experimental observations, to measures, definitions, equipment, theories, and more. Science is reliable in large part because it is a messy network of interconnected endeavours, of diverse types, that draw from a wide variety of sources. To claim that you can rely on the general theory of relativity to build satellites, the taxonomy of

Closing Figure An illustration of floating nests from *The Universe, Or, The Wonders of Creation: The Infinitely Great and the Infinitely Little* by F.A. Pouchet (1883). Illustration by A. Mesnel
Source: Classic Image/Alamy Stock Photo.

birds to classify a new species, the success of cancer treatment to cure a patient, or the theory of the democratic peace to predict the outcome of the 'cod wars' without acknowledging that this reliance is, in turn, backed by a virtuous tangle of work that has been built by generations of scientists is to ignore the way science works. Good science is inescapably rich, entangled, and long-tailed.

References

Abbott, B.P., Abbott, R., Abbott, T.D., Abernathy, M.R., Acernese, F., Ackley, K., Adams, C., Adams, T., Addesso, P., Adhikari, R.X., and Adya, V.B. (2016), 'Observation of gravitational waves from a binary black hole merger', *Physical Review Letters* 116/6: 061102.

Abraham, E.P. and Chain, E. (1940), 'An enzyme from bacteria able to destroy penicillin', *Nature* 146/837: 837.

Abraham, E.P. et al. (1941), 'Further observations on penicillin', *The Lancet* 238/6155: 177–89.

Andersen, H. and Hepburn, B. (2015), 'Scientific method', in Edward N. Zalta (ed.), *The Stanford Encyclopedia of Philosophy*. https://plato.stanford.edu/archives/win2020/entries/scientific-method/.

Angrist, J.D. and Pischke, J.-S. (2010), 'The credibility revolution in empirical economics: how better research design is taking the con out of econometrics', *Journal of Economic Perspectives* 24/2: 3–30.

Atkinson, A.B. (1998), *Poverty in Europe: Jrjo Jahnsson Lectures* (Wiley-Blackwell).

Austin, J.L. (1962), *Sense and Sensibilia* (Oxford University Press).

Barlow, S.P. (2017), '5 Surprising uses for the Swiss army knife', Swiss Army Survival, August 15, Magazine: *Recoil Offgrid*. https://www.offgridweb.com/gear/5-surprising-uses-for-the-swiss-army-knife/.

Barrotta, P. and Montuschi, E. (2018a), 'The dam project: who are the experts? Philosophical lessons from the Vajont disaster', in Barrotta, P. and Scarafile, G. (eds), *Science and Democracy: Controversies and Conflicts* (J. Benjamin).

Barrotta, P. and Montuschi, E. (2018b), 'Expertise, relevance and types of knowledge: social epistemology', *Social Epistemology* 32/6: 387–96.

Batterman, R.W. (2009), 'Idealization and modeling', *Synthese* 169/3: 427–46.

Ben-David, J. and Collins, R. (1966), 'Social factors in the origins of a new science', *American Sociological Review*, 31/4: 451–65.

Bennett, A. and Checkel, J.T. (2015), *Process Tracing* (Cambridge University Press).

Bennett, K. and McLaughlin, B. ([2005] 2018), 'Supervenience', in Edward N. Zalta (ed.), *The Stanford Encyclopedia of* Philosophy. https://plato.stanford.edu/archives/win2018/entries/supervenience/.

Bhakthavatsalam, S. (2019), 'The value of false theories in science education', *Science and Education* 28: 5–23.

Bhakthavatsalam, S. and Cartwright, N. (2017), 'What's so special about empirical adequacy?', *European Journal for Philosophy of Science* 7/3: 445–65.

Bod, R., van Dongen, J., ten Hagen, S.L., Karstens, B., and Mojet, E. (2019), 'The flow of cognitive goods: a historiographical framework for the study of epistemic transfer', *Isis* 110/3: 483–96.

Boehner, P. (1964), *William of Ockham, Philosophical Writings*, The Library of Liberal Arts, fourth printing edition (Bobbs-Merrill Company).

Boese, V.A. (2019), 'How (not) to measure democracy', *International Area Studies Review* 22/2: 95–127.

Bolinska, A. and Martin, J.D. (2020), 'Negotiating history: contingency, canonicity, and case studies', *Studies in History and Philosophy of Science Part A* 80: 37–46.

Bourne, J., Donnelly, C.A., Cox, D.R., Gettinby, G., McInerney, J.P., Morrison, W.I., and Woodroffe, R. (2007), *Bovine TB: The Scientific Evidence. The Final Report of the Independent Scientific Group on Cattle TB* (London: Department of Environment, Food and Rural Affairs).

Bradford-Hill, A. (1965), 'The environment and disease: association or causation?', *Proceedings of the Royal Society of Medicine* 58/5: 295–300.

Brandom, R. (1998), *Making It Explicit: Reasoning, Representing, and Discursive Commitment* (Harvard University Press).

Brown, M.J. (2019), 'Is science really value free? From objectivity to scientific integrity', in McCain, K. and Kampourakis, K. (eds), *What Is Scientific Knowledge? An Introduction to Contemporary Epistemology of Science*, chapter 15 (Routledge).

Bud, R. (2007), *Penicillin: Triumph and Tragedy* (Oxford University Press).

Burch, M. and Furman, K. (2019), 'Objectivity in science and law: a shared rescue strategy', *International Journal of Law and Psychiatry* 64: 60–70.

Bush, G.W. (2014), 'President and Prime Minister Blair discussed Iraq, Middle East', 12 November. https://georgewbush-whitehouse.archives.gov/news/releases/2004/11/20041112-5.html.

Cairns, B.H., Eden, K.C., and Shoreston, J. (1944), 'A review of the Florey and Cairns Report on the use of penicillin in war wounds', *Journal of Neurosurgery* 1: 201–10.

Caporael, L.R., Griesemer, J.R., and Wimsatt, W.C. (eds) (2014), *Developing Scaffolds in Evolution, Culture, and Cognition* (MIT Press).

Cardin, J.A., Carlén, M.K., Knoblich, U., Zhang, F., Deisseroth, K., Tsai, L., and Moore, C. (2010), 'Targeted optogenetic stimulation and recording of neurons in vivo using cell-type-specific expression of Channelrhodopsin-2', *Nature Protocols* 5/2: 247–54.

Carloni, G.C. (1995), *Il Vaiont trent'anni dopo: esperienza di un geologo* (Clueb).

Cartwright, N. (1999), *The Dappled World: A Study of the Boundaries of Science* (Cambridge University Press).

Cartwright, N. (2014), 'A question of nonsense', *Iyyun: The Jerusalem Philosophical Quarterly* 63/January: 102–16.

Cartwright, N. (2019), *Nature, The Artful Modeler: Lectures on Laws, Science, How Nature Arranges the World and How We Can Arrange It Better*, The Paul Carus Lecture Series, Vol. 23 (Open Court Publishing).

Cartwright, N. (2020), 'Why trust science? Reliability, particularity and the tangle of science', *Proceedings of the Aristotelian Society* 120/3: 237–52.

Cartwright, N. and Bradburn, N. (2011), *A Theory of Measurement* (National Academies Press).

Cartwright, N., Charlton, L., Juden, M., Munslow, T., and Williams, R.B. (2020), 'Making predictions of programme success more reliable', CEDIL Methods Working Paper 1 (Oxford: Centre of Excellence for Development Impact and Learning).

Cartwright, N. and Hardie, J. (2012), *Evidence-Based Policy: A Practical Guide to Doing It Better* (Oxford University Press).

Cartwright, N. and Montuschi, E. (2014), *Philosophy of Social Science: A New Introduction* (Oxford University Press).

Cavendish, H. (1776), 'XXI. An account of the meteorological instruments used at the Royal Society's house', *Philosophical Transactions of the Royal Society of London* 66: 375–401.

Cavendish, H., Heberden, W., Aubert, A., Luc, J.A.D., Maskelyne, N., Horsley, S., and Planta, J. (1777), 'XXXVII. The report of the Committee appointed by the Royal Society to consider of the best method of adjusting the fixed points of thermometers; and of the precautions necessary to be used in making experiments with those instruments', *Philosophical Transactions of the Royal Society of London* 67: 816–57.

Cetina, K.K. (2009), *Epistemic Cultures: How the Sciences Make Knowledge* (Harvard University Press).

Chain, E., Florey, H.W., Gardner, A.D., Heatley, N.G., Jennings, M.A., Orr-Ewing, J., and Sanders, A.G. (1940), 'Penicillin as a chemotherapeutic agent', *The Lancet* 236/6104: 226–8.

Chamberlain, T. (2020), 'The capabilities approach to well-being: characterizing capabilities and measuring them' (dissertation, University of California, San Diego).

Chang, H. (2004), *Inventing Temperature: Measurement and Scientific Progress* (Oxford University Press).

Chang, H. (2012), *Is Water H₂O?: Evidence, Realism and Pluralism*, Boston Studies in the Philosophy of Science, Vol. 293 (Springer Science & Business Media).

Chang, H. (2017), 'July. VI—operational coherence as the source of truth', *Proceedings of the Aristotelian Society* 117/2: 103–22.

Chapman, R. and Wylie, A. (2016), *Evidential Reasoning in Archaeology* (Bloomsbury).

Clutterbuck, P.W., Lovell, R., and Raistrick, H. (1932), 'Studies in the biochemistry of micro-organisms: the formation from glucose by members of the penicillium chrysogenum series of a pigment, an alkali-soluble protein and penicillin—the antibacterial substance of Fleming', *Biochemical Journal* 26/6: 1907–18.

Coll, R.K., Chang, W.H., Dillon, J., Justi, R., Mortimer, E., Tan, K.C.D., Treagust, D.F., and Paul, W. (2009), 'An international perspective of monitoring educational research quality: commonalities and differences', in Shelley, M., Yore, L., and Hand, B. (eds), *Quality Research in Literacy and Science Education*, 107–37 (Springer).

Collins, P.H. (1986), 'Learning from the outsider within: the sociological significance of black feminist thought', *Social Problems* 33/6: 14–32.

Conway, E.C. and Oreskes, N. (2011), *Merchants of Doubt* (Bloomsbury).

Coppedge, M., Gerring, J., Lindberg, S.I., Skaaning, S.E., and Teorell, J. (2017), 'V-Dem comparisons and contrasts with other measurement projects', V-Dem Working Paper 2017/45. https://www.v-dem.net/media/publications/v-dem_working_paper_2017_45.pdf.

Crasnow, S. (2013), 'Feminist philosophy of science: values and objectivity', *Philosophy Compass* 8/4, 413–23.

Crasnow, S. (2021), 'Coherence objectivity and measurement: the example of democracy', *Synthese* 199: 1207–29.

Creighton, J.D. and Anderson, W.G. (2012), *Gravitational-Wave Physics and Astronomy: An Introduction to Theory, Experiment and Data Analysis* (John Wiley & Sons).

Crombie, A.C. (1953), *Robert Grosseteste and the Origins of Experimental Science, 1100–1700* (Clarendon Press).

Crombie, A.C. (1994), *Styles of Scientific Thinking in the European Tradition: The History of Argument and Explanation Especially in the Mathematical and Biomedical Sciences and Arts*, Vol. 2 (Duckworth).

Dahl, R.A. (1991), *Democracy and Its Critics* (Yale University Press).

Daston, L. and Galison, P. (2007), *Objectivity* (Zone Books).

Davey, C.H., Bonell, C., and Cartwright, N. (2017), 'Gaps in evaluation methods for addressing challenging contexts in development', CEDIL pre-inception paper. https://doi.org/10.13140/RG.2.2.18900.86400.

Davidson, D. (1992), 'Thinking causes', in Heil, J. and Mele, A.R. (eds), *Mental Causation*, 3–18 (Oxford University Press).

Davidson, D. (1995), 'Laws and cause', *Dialectica* 49: 263–80.

Deaton, A. and Cartwright, N. (2018), 'Understanding and misunderstanding randomized controlled trials', *Social Science & Medicine* 210: 2–21.

Di Bucchianico, M.E. (2009), 'Modelling high-temperature superconductivity: a philosophical inquiry in theory, experiment and dissent' (doctoral dissertation, London School of Economics and Political Science).

Douglas, H. (2000), 'Inductive risk and values in science', *Philosophy of Science* 67/4: 559–79.

Douglas, H. (2004), 'The irreducible complexity of objectivity', *Synthese* 138/3: 453–73.

Drake, S. (1972), *Galileo Studies: Personality, Tradition, and Revolution* (Cambridge University Press).

Drake, S. (1973), 'Galileo's discovery of the law of free fall', *Scientific American* 228/5: 84–93.

Düerr, P.M. (2019), 'Fantastic beasts and where (not) to find them: local gravitational energy and energy conservation in general relativity', *Studies in History and Philosophy of Science Part B: Studies in History and Philosophy of Modern Physics* 65: 1–14.

Duflo, E. and Kremer, M. (2003), 'Use of randomization in the evaluation of development effectiveness', Paper Prepared for the World Bank Operations Evaluation Department (OED) Conference on Evaluation and Development Effectiveness in Washington, DC, 15–16 July 2003.

Duhem, P. ([1906] 1982), *The Aim and Structure of Physical Theory* (Princeton University Press).

Dupré, J. (1995), *The Disorder of Things: Metaphysical Foundations of the Disunity of Science* (Harvard University Press).

Einstein, A. (1933), 'On the method of theoretical physics: the Herbert Spencer lecture', Philosophy of Science, 1/2 (1934): 163–9.

Einstein, A. (1997), *The Collected Papers of Albert Einstein, Volume 6: The Berlin Years: Writings, 1914–1917: English Translation of Selected Texts* (Princeton University Press).

Eisenhardt, K.M., Graebner, M.E., and Sonenshein, S. (2016), 'From the editors. Grand challenges and inductive methods: rigor without rigor mortis', *Academy of Management Journal* 59/4: 1113–23.

Elgin, C. (2009), 'Exemplification, idealization, and scientific understanding', in Suárez, M. (ed.), *Fictions in Science: Essays on Idealization and Modeling*, 77–90 (Routledge).

Elster, J. (2015), *Explaining Social Behavior: More Nuts and Bolts for the Social Sciences* (Cambridge University Press).

Eronen, M.I. (2019), 'Psychopathology and truth: a defense of realism', *The Journal of Medicine and Philosophy: A Forum for Bioethics and Philosophy of Medicine* 44/4: 507–20.

Faget, L., Zell, V., Souter, E., McPherson, A., Ressler, R., Gutierrez-Reed, N., Yoo, J.H., Dulcis, D., and Hnasko, T.S. (2018), 'Opponent control of behavioral reinforcement by inhibitory and excitatory projections from the ventral pallidum', *Nature Communications* 9/849, 1–14.

Feyerabend, P. (1962), 'Explanation, reduction, and empiricism', in Feigl, H. and Maxwell, G. (eds), *Scientific Explanation, Space, and Time*, 28–97, Minnesota Studies in the Philosophy of Science, Vol. 3 (Minnesota University Press).

Feyerabend, P. (1975), *Against Method: Outline of an Anarchistic Theory of Knowledge* (Verso).

Fine, A. (1998), 'The viewpoint of no-one in particular', *Proceedings and Addresses of the American Philosophical Association* 72/2: 7–20.

Fleck, L. ([1935] 1979), *Genesis and Development of a Scientific Fact* (University of Chicago Press).

Fleming, A. (1922), 'On a remarkable bacteriolytic element found in tissues and secretions', *Proceedings of the Royal Society B: Biological Sciences* 93/653: 306–17.

Fleming, A. (1929), 'On the antibacterial action of cultures of a penicillium, with special reference to their use in the isolation of B. influenzae', *British Journal of Experimental Pathology* 3: 226–36.

Fleming, A. (1932), 'On the specific antibacterial properties of penicillin and potassium tellurite: incorporating a method demonstrating some bacterial antagonisms', *The Journal of Pathology and Bacteriology* 35/6: 831–42.

Florey, H.W. (1946), 'Penicillin (a lecture)', *Experientia* 2: 160–71.

Florey, M.E., Adelaide, M.B., and Florey, H.W. (1943), 'General and local administration of penicillin', *The Lancet* 241/6239: 387–97.

Galilei, G. ([1632] 2001), *Dialogue Concerning the Two Chief World Systems, Ptolemaic and Copernican* (Modern Library Classics).

Galison, P. (1997), *Image and Logic: A Material Culture of Microphysics* (University of Chicago Press).

Gallin, J., Ognibene, F., and Johnson, L. (2017), *Principles and Practice of Clinical Research* (Elsevier).

Gauch, Jr, H.G. and Gauch, H.G. (2003), *Scientific Method in Practice* (Cambridge University Press).

Gervasoni, A. (1969), *Il Vajont e le responsabilità dei manager* (Bramante editrice).

Gimbel, S. (2011), *Exploring the Scientific Method: Cases and Questions* (University of Chicago Press).

Gollub, J.P. (2003), 'Discrete and continuum descriptions of matter', *Physics Today* 56/1, https://doi.org/10.1063/1.1554120.

Gower, B. (1997), *Scientific Method: An Historical and Philosophical Introduction* (Psychology Press).

Graney, C.M. (2015), *Setting Aside All Authority* (University of Notre Dame Press).

Guba, E.G. and Lincoln, Y.S. (1989), *Fourth Generation Evaluation* (Sage).

Hacking, I. (1994), 'Styles of scientific thinking or reasoning: a new analytical tool for historians and philosophers of the sciences', in Cohen, R.S. (ed.), *Trends in the Historiography of Science*, 31–48 (Boston University Press).

Hacking, I. (2015), 'Let's not talk about objectivity', in Tsou, J.Y., Richardson, A., and Padovani, F. (eds), *Objectivity in Science*, 19–33 (Springer Verlag).

Haidich, A.B. (2010), 'Meta-analysis in medical research', *Hippokratia* 14 (Suppl. 1): 29–37.

Hanson, N.R. (1958), *Patterns of Discovery: An Inquiry into the Conceptual Foundations of Science* (Cambridge University Press).

Harding, S. (1991), *Whose Science? Whose Knowledge?* (Cornell University Press).

Harding, S.G. (2004), *The Feminist Standpoint Theory Reader: Intellectual and Political Controversies* (Psychology Press).

Hempel, C.G. (1952), *Fundamentals of Concept Formation in Empirical Science*, Vol. 2 (University of Chicago Press).

Hempel, C.G. (1988), 'Provisoes: a problem concerning the inferential function of scientific theories', *Erkenntnis* 28(2): 147–64.

Hendry, D. (1980), 'Econometrics-alchemy or science?', *Economica* 47/188: 387–406.

Hesse, M. ([1963] 1966), *Models and Analogies in Science* (Sheed and Ward; revised edn University of Notre Dame Press).

Hitchcock, C. and Sober, E. (2004), 'Prediction versus accommodation and the risk of overfitting', *The British Journal for the Philosophy of Science* 55/1: 1–34.

Hoefer, C. (2000). 'Energy conservation in GTR', *Studies in History and Philosophy of Science Part B: Studies in History and Philosophy of Modern Physics* 31/2: 187–99.

Howlett, P. and Morgan, M. (2010), *How Facts Travel* (Cambridge University Press).

Jones, B., Jarvis, P., Lewis, J.A., and Ebbutt, A.F. (1996), 'Trials to assess equivalence: the importance of rigorous methods', *British Medical Journal* 313/7048: 36–9.

Kant, I. ([1795] 2020), *Perpetual Peace: A Philosophical Text* (BoD [Books on Demand]).

Keas, M.N. (2018), 'Systematizing the theoretical virtues', *Synthese* 195/6: 2761–93.

King, D., Roper, T.J., Young, D., Woolhouse, M.E.J., Collins, D.A., and Wood, P. (2007), *Bovine tuberculosis in cattle and badgers: a report by the Chief Scientific Adviser, Sir David King* (Defra).

Kitcher, P. (1993), *The Advancement of Science* (Oxford University Press).

Kitcher, P. (2003), *Science, Truth, and Democracy* (Oxford University Press).

Koskinen, I. (2020), 'Defending a risk account of scientific objectivity', *The British Journal for the Philosophy of Science* 71/4: 1187–207.

Koskinen, I. (2021), 'Objectivity in contexts: withholding epistemic judgement as a strategy for mitigating collective bias', *Synthese* 199: 211–25.

Kuhn, Thomas (1973), 'Objectivity, value judgment, and theory choice', in *The Essential Tension: Selected Studies in Scientific Tradition and Change* (University of Chicago Press).

Kuhn, T.S. (1962), *The Structure of Scientific Revolutions* (University of Chicago Press).

Kuhn, T.S. (1977), 'Objectivity, value judgment, and theory choice', in Bird, A. and Ladyman, J. (eds), *Arguing About Science*, 74–86 (Routledge).

Kukla, A. (1996), 'Does every theory have empirically equivalent rivals?', *Erkenntnis* 44/2: 137–66.

Lange, M. (2002), *An Introduction to the Philosophy of Physics: Locality Fields, Energy, and Mass* (John Wiley and Sons).

Latour, B. (2005), *Reassembling the Social: An Introduction to Actor-Network Theory* (Oxford University Press).

Laudan, L. (1981), 'A confutation of convergent realism', *Philosophy of Science* 48/1: 19–49.

Laudan, L. (1984), 'Reconstructing methodology', in Anderson, P.F. and Ryan, M.J. (eds), *Scientific Method in Marketing: Winter Educators' Conference*, 1–4 (American Marketing Association).

Laudan, L. (2004), 'The epistemic, the cognitive, and the social', in Machamer P. and Wolters G. (eds), *Science, Values, and Objectivity*, 14–23 (University of Pittsburgh Press).

Laudan, L. and Leplin, J. (1991), 'Empirical equivalence and underdetermination', *Journal of Philosophy* 88/9: 449–72.

Leamer, E.E. (1983), 'Let's take the con out of econometrics', *The American Economic Review* 73/1: 31–43.

Leonelli, S. (2016), *Data-Centric Biology. A Philosophical Study* (University of Chicago Press).

Leonelli, S. and Ankeny, R. (2015), 'Repertoires: how to transform a project into a research community', *Bioscience* 65/7: 701–8.

Leplin, J. (1997), *A Novel Defense of Scientific Realism* (Oxford University Press).

Lewis, P.J. (2001), 'Why the pessimistic induction is a fallacy', *Synthese* 129: 371–80.

Li, B. (2019), 'Central amygdala cells for learning and expressing aversive emotional memories', *Current Opinion in Behavioral Sciences* 26: 40–5.

Lincoln, Y.S. and Guba, E.G. (2007), 'Naturalistic inquiry', in Ritzer, G. (ed.), *The Blackwell Encyclopedia of Sociology* (Blackwell).

Lloyd, E.A. (1996), 'Science and anti-science: objectivity and its real enemies', in Nelson, J. (ed.), *Feminism, Science, and the Philosophy of Science*, 217–59 (Springer).

Longino, H.E. (1990), *Science as Social Knowledge: Values and Objectivity in Scientific Inquiry* (Princeton University Press).

Longino, H.E. (1996), 'Cognitive and non-cognitive values in science: rethinking the dichotomy', in Nelson, J. (ed.), *Feminism, Science, and the Philosophy of Science*, 39–58 (Springer).

Longino, H.E. (2001), 'What do we measure when we measure aggression?', *Studies in History and Philosophy of Science Part A* 32/4: 685–704.

Longino, H.E. (2002), *The Fate of Knowledge* (Princeton University Press).

Luce, R.D. and Suppes, P. (2002), 'Representational measurement theory', in Pashler, H. and Wixted, J. (eds), *Stevens' Handbook of Experimental Psychology*, 1–41 (John Wiley & Sons).

McKeon, R. (1941), 'Posterior analytics', in *The Basic Works of Aristotle* (Random House).

McMullin, E. (1967), *Galileo: Man of Science* (Basic Books).

McTighe, T.P. (1967), 'Galileo's Platonism: a reconsideration', in McMullin, E., *Galileo: Man of Science* (Basic Books).

Maudlin, T. (2011), *Quantum non-locality and relativity: metaphysical intimations of modern physics* (John Wiley & Sons).

Megill, A. (1994), *Rethinking Objectivity* (Duke University Press).

Middleton, W.E. Knowles (1966), *A History of the Thermometer and Its Uses in Meteorology* (Johns Hopkins University Press).

Mill, J.S. ([1836] 1967), 'On the definition of political economy and the method of investigation proper to it'. Reprinted in Mill, J.S. and Robson, J.M. (eds), *Collected Works of John Stuart Mill: Essays on Economics and Society, 1* (University of Toronto Press).

Montuschi, E. (2014), 'Scientific objectivity', in Cartwright, N. and Montuschi, E. (eds), *Philosophy of Social Science: A New Introduction*, 123–44 (Oxford University Press).

Montuschi, E. (2016), 'Objectivity', in McIntyre, L. and Rosenebrg, A. (eds), *The Routledge Companion in the Philosophy of Social Science*, 281–91 (Routledge).

Montuschi, E. (2017a), 'Using science, making policy: what should we worry about?', *European Journal for Philosophy of Science* 7/1: 57–78.

Montuschi, E. (2017b), 'Scientific evidence vs. expert opinion: a false alternative?', *Notizie di Politeia* 32/12: 60–79.

Montuschi, E. (2021), 'Finding a context for objectivity', *Synthese* 199: 4061–76.

Morales, M. and Margolis, E.B. (2017), 'Ventral tegmental area: cellular heterogeneity, connectivity, and behaviour', *Nature Reviews Neuroscience* 18/2: 73–85.

Morgan, M. and Morrison, M. (1999), *Models as Mediators* (Cambridge University Press).

Müller, H. and Wolff, J. (2006), 'Democratic peace: many data, little explanation?', in Geis, A., Brock, L., and Mueller, H. (eds), *Democratic Wars*, 41–73 (Palgrave Macmillan).

Munro, E. and Hardie, J. (2019), 'Why we should stop talking about objectivity and subjectivity in social work', *The British Journal of Social Work* 49/2: 411–27.

Munslow, T. (n.d.), 'Values in science and technology: ethical considerations in the evaluation of undernutrition programmes' (unpublished doctoral thesis, Durham University).

Murad, M.H., Asi, N., Alsawas, M., and Alahdab, F. (2016), 'New evidence pyramid', *BMJ Evidence-Based Medicine* 21/4: 125–7.

Nagel, E. (1961), 'The value-oriented bias of social enquiry', in *The Structure of Science: Problems in the Logic of Scientific Explanation* (Hackett).

Nagel, T. (1986), *The View from Nowhere* (Oxford University Press).

Neurath, O. ([1921] 1973), 'Anti-Spengler', in Neurath, M. and Cohen, R.S. (eds), *Empiricism and Sociology*, 158–213 (D. Reidel).

Neurath, O. (1983), 'Physicalism and the investigation of knowledge', in Cohen, R.S. and Neurath, M. (eds), *Philosophical Papers 1913–1946*, 159–71 (Springer).

Newton, I. ([1687] 1999), *The Principia: Mathematical Principles of Natural Philosophy: A New Translation*, ed. Cohen, I.B. and Whitman, A. (University of California Press).

Norton, J.D. (2003a), 'Causation as folk science', *Philosophers' Imprint* 3/4. www.philosophersimprint.org/003004/.

Norton, J.D. (2003b), 'A material theory of induction', *Philosophy of Science* 70/4: 647–70.

Nozick, R. (1998), 'Invariance and objectivity', *Proceedings and Addresses of the American Philosophical Association* 72/2: 21–48.

Nussbaum, M. and Sen, A. (1993), *The Quality of Life* (Clarendon Press).

Ockham, W. (1964), *Ockham's Philosophical Writings: A Selection*, ed. and trans. Boehner, P., revised Brown, S.F. (Hackett).

Olds, J. (1969), 'The central nervous system and the reinforcement of behavior', *American Psychology* 24/2: 114–32.

Olds, J. and Milner, P. (1954), 'Positive reinforcement produced by electrical stimulation of septal area and other regions of rat brain', *Journal of Comparative and Physiological Psychology* 47/6: 419–27.

Oreskes, N. (2019), *Why Trust Science?* (Princeton University Press).

Pawson, R. and Tilley, N. (1997), *Realistic Evaluation* (Sage).

Pera, M. (1994), *The Discourses of Science* (University of Chicago Press).

Pickering, A. (1995), *The Mangle of Practice: Time, Agency, Science* (University of Chicago Press).

Pirani, F.A. (1957), 'Invariant formulation of gravitational radiation theory', *Physical Review* 105/3: 1089.

Popay, J. (2006), *Moving Beyond Effectiveness in Evidence Synthesis: Methodological Issues in the Synthesis of Diverse Sources of Evidence* (National Institute for Health and Clinical Excellence).

Popper, K. (2015), 'The logic of scientific discovery', in Shand, J. (ed.) *Central Works of Philosophy. Volume 4: The Twentieth Century: Moore to Popper* (McGill-Queen's University Press).

Popper, K.R. ([1935] 1963), 'Science as falsification', *Conjectures and Refutations* 1: 33–9.

Porter, T.M. ([1995] 2020), *Trust in Numbers: The Pursuit of Objectivity in Science and Public Life* (Princeton University Press).

Pouchet, F.-A. (1883), *The Universe, Or, The Wonders of Creation: The Infinitely Great and the Infinitely Little* (H. Hallett).

Pritle, Z. (2020), 'Unity of engineering disciplines', *The Bridge*, Winter: 67–9.

Putnam, H. (1981), *Reason, Truth and History*, Vol. 3 (Cambridge University Press).

Quine, W.V.O. (1951), 'Two dogmas of empiricism', *Philosophical Review* 60/1: 20–43.

Ray, J.L. (1993), 'Wars between democracies: rare, or nonexistent?', *International Interactions* 18/3: 251–76.

Read, J. (2020), 'Functional gravitational energy', *The British Journal for the Philosophy of Science* 71/1: 205–32.

Reiss, J. (2013), *Philosophy of Economics: A Contemporary Introduction* (Routledge).

Reiss, J. ([2008] 2016), *Error in Economics: Towards a More Evidence–Based Methodology* (Routledge).

Reiss, J. and Sprenger, J. (2014), 'Scientific objectivity', in Edward N. Zalta (ed.), *The Stanford Encyclopedia of Philosophy*. https://plato.stanford.edu/entries/scientific-objectivity/.

Risse-Kappen, T. (1995), 'Democratic peace—warlike democracies? A social constructivist interpretation of the liberal argument', *European Journal of International Relations* 1/4: 491–517.

Ross, J.F. (1981), *Portraying Analogy* (Cambridge University Press).

Rudner, R. (1953), 'The scientist qua scientist makes value judgments', *Philosophy of Science* 20/1: 1–6.

Russett, B. (1994), *Grasping the Democratic Peace: Principles for a Post-Cold War World* (Princeton University Press).

Schattschneider, E.E. (1960), *Party Government* (Transaction).

Schindler, S. (2018), *Theoretical Virtues in Science: Uncovering Reality through Theory* (Cambridge University Press).

Schmitz, C. and Hof, P.R. (2000), 'Recommendations for straightforward and rigorous methods of counting neurons based on a computer simulation approach', *Journal of Chemical Neuroanatomy* 20/1: 93–114.

Schulmann, R., Kox, A.J., Janssen, M., and Illy, J. ([1916] 1998), *The Collected Papers of Albert Einstein, Volume 8: The Berlin Years: Correspondence 1914–1918* (Princeton University Press).

Seale, C. (1999), 'Quality in qualitative research', *Qualitative Inquiry* 5/4, 465–78.

Sellars, W. (1968), *Science and Metaphysics: Variations on Kantian Themes* (Routledge & Kegan Paul and The Humanities Press).

Settle, T.B. (1967), 'Galileo's use of experiment as a tool of investigation', in McMullin, E. (ed.), *Galileo: Man of Science* (Basic Books).

Shapere, D. (1974), *Galileo: A Philosophical Study* (University of Chicago Press).

Short, T.G. and Leslie, K. (2014), '"Known unknowns and unknown unknowns": electroen-cephalographic burst suppression and mortality', *British Journal of Anaesthesia* 113/6: 897–9.

Sibbald, B. and Roland, M. (1998), 'Understanding controlled trials: why are randomised con-trolled trials important?', *British Medical Journal* 316/7126: 201.

Sims, C.A. (1980), 'Macroeconomics and reality', *Econometrica* 48(1): 1–48.

Skeem, J.L. and Lowenkamp, C.T. (2016), 'Risk, race, and recidivism: predictive bias and dis-parate impact', *Criminology* 54/4: 680–712.

Sklar, L. (1981), 'Do unborn hypotheses have rights?', *Pacific Philosophical Quarterly* 62: 17–29.

Sober, E. (2015), 'Is the scientific method a myth? Perspectives from the history and philoso-phy of science', *METODE Science Studies Journal* 5: 195–9.

Soler, L., Zwart, S., Lynch, M., Israel-Jost, V. (2014), *Science After the Practice Turn in the Philosophy, History, and Social Studies of Science* (New York/London: Routledge).

Spencer, L., Ritchie, J., Lewis, J., and Dillon, L. (2003), *Quality in Qualitative Evaluation: A Framework for Assessing Research Evidence* (Government Chief Social Researcher's Office).

Stanford, P.K. (2006), *Exceeding Our Grasp: Science, History, and the Problem of Unconceived Alternatives*, Vol. 1 (Oxford University Press).

Stewart, P. (1964), 'Jacobellis v Ohio', *US Rep* 378: 184.

Stiglitz, J.E., Sen, A., and Fitoussi, J.P. (2009), 'Report by the commission on the measurement of economic performance and social progress'. https://www.economie.gouv.fr/files/finances/presse/dossiers_de_presse/090914mesure_perf_eco_progres_social/synthese_ang.pdf.

Strevens, M. (2012), 'Ceteris Paribus hedges: causal voodoo that works', *Journal of Philosophy* 109: 652–75.

Strevens, M. (2020), *The Knowledge Machine: How Irrationality Created Modern Science* (Allen Lane).

Tichy, N.M. and Bennis, W.G. (2007), 'Making judgment calls', *Harvard Business Review* 85/10: 94.

Towne, L. and Shavelson, R.J. (2002), *Scientific Research in Education* (National Academy Press Publications Sales Office).

Trueta, J. (1980), *Trueta, Surgeon in War and Peace: The Memoirs of Josep Trueta, translated by Amelia and Michael Strubell* (Victor Gollancz).

Tschauner, H. (1996), 'Middle-range theory, behavioral archaeology, and postempiricist philosophy of science in archaeology', *Journal of Archaeological Method and Theory* 3/1: 1–30.

Ullmann-Margalit, E. (2006), *Out of the Cave: A Philosophical Inquiry into the Dead Sea Scrolls Research* (Harvard University Press).

US National Research Council (2002), *Scientific Research in Education* (The National Academies Press).

Van Evera, S. ([1997] 2015), *Guide to Methods for Students of Political Science* (Cornell University Press).

Vickers, A.J. (2006), 'How to randomize', *Journal of the Society for Integrative Oncology* 4/4: 194.

Vickers, P. (2022), *Identifying Future-Proof Science* (Oxford University Press).

Weber, M. (1949), 'Objectivity in social science and social policy', in *Max Weber on the Methodology of the Social Sciences*, 50–112, trans. and ed. Shils, E.A. and Finch, H.A. (The Free Press).

Whelan, F.G. (1983), 'Prologue: democratic theory and the boundary problem', *Nomos* 25: 13–47.

White, H. (2009), 'Theory-based impact evaluation: principles and practice', *Journal of Development Effectiveness* 1/3: 271–84.

White, H. and Masset, E. (2007), 'The Bangladesh integrated nutrition program: findings from an impact evaluation', *Journal of International Development*, 19: 627–52.

White, S.C. (1992), *Arguing with the Crocodile: Gender and Class in Bangladesh* (Zed Books).

Williams, B. (1986), *Ethics and the Limits of Philosophy* (Fontana).

Wilson, T.D., Valdivia, S., Khan, A., Ahn, H., Adke, A.P., Gonzalez, S.M., Sugimura, Y.K., and Carrasquillo, Y. (2019), 'Dual and opposing functions of the central amygdala in the modulation of pain', *Cell Reports* 29/2: 332–46.

Wilson, W. (1917), 'Joint address to Congress leading to a Declaration of War against Germany (1917)'. https://www.archives.gov/milestone-documents/address-to-congress-declaration-of-war-against-germany.

Wittgenstein, L. (1953), *Philosophical Investigations*, trans. Anscombe, G.E.M. (Blackwell).

Woody, A. (2015), 'Re-orienting discussions of scientific explanation: a functional perspective', *Studies in History and Philosophy of Science Part A* 52: 79–87.

World Bank (1995), *Staff Appraisal Report: Bangladesh Integrated Nutrition Project* (World Bank Population and Human Resources Division).

Worsdale, R. and Wright, J. (2021), 'My objectivity is better than yours: contextualising debates about gender inequality', *Synthese*, 199: 1659–83.

Wray, K.B. (2013), 'Success and truth in the realism/anti-realism debate', *Synthese*, 190: 1719–29.

Wright, J. (2018), 'Rescuing objectivity: a contextualist proposal', *Philosophy of the Social Sciences*, 48/4: 385–406.

Wu, Z.Y. (2016), 'Gravitational energy-momentum and conservation of energy-momentum in general relativity', *Communications in Theoretical Physics* 65/6: 716.

Wylie, A. (1989a), 'Archaeological cables and tacking: the implications of practice for Bernstein's "options beyond objectivism and relativism"', *Philosophy of the Social Sciences* 19/1: 1–18.

Wylie, A. (1989b), 'Matters of fact and matters of interest', in Shennan, S. (ed.), *Archaeological Approaches to Cultural Identity*, 94–109 (Unwin Hyman).

Wylie, A. (2003), 'Why standpoint matters. Science and other cultures: issues', *Philosophies of Science and Technology* 26: 48.

Yang, L.J., Chang, K.W., and Chung, K.C. (2012), 'Methodologically rigorous clinical research', *Plastic and Reconstructive Surgery*, 129/6: 979e–88e.

Index

Note: Tables and figures are indicated by an italic '*t*' and '*f*', respectively, following the page number.

For the benefit of digital users, indexed terms that span two pages (e.g., 52–53) may, on occasion, appear on only one of those pages.